Gas Works, Near the Regent's Canal, London Circa 1828

Natural Gas Fundamentals

Malcolm W.H. Peebles

Shell International Gas Limited

© Shell International Gas Limited 1992

All rights reserved. No part of this publication may be reproduced or transmitted, in any form or by any means, without permission.

First published 1992 by
SHELL INTERNATIONAL GAS LIMITED, Shell Centre, London SE1 7NA, England

Printed in Bath, England

British Library Cataloguing in Publication Data

Peebles, Malcolm W.H.
    Natural Gas Fundamentals. (hdbk).
    ISBN 0-9519299-0-9

Produced by Shell Publicity Services ODLS April 1992 Job No. 90156 5m   Printed in England by Bath Press

Contents

	Page No.
List of Figures	vi
List of Tables	x
Preface	xi
About the Book	xii
About the Author	xiii
Acknowledgements	xiii

1: WHAT IS NATURAL GAS? — 1
- Associated and non-associated natural gas — 1
- Composition and terminology — 1
- Quality considerations — 2
- Gas treatment — 5
- Combustion characteristics — 5

2: EXPLORATION — 9
- Some basic geological facts — 9
- Oil and gas accumulations — 12
- Gas traps — 12
- Exploration methods — 13
- Seismic — 14
- Exploration drilling — 15
- Appraisal drilling and commerciality considerations — 17
- Fiscal framework — 18
- Footnotes — 19

3: PRODUCTION — 21
- Natural gas reserves — 21
- Pre-development work — 22
- The development phase — 24
- Offshore production platforms — 25
- The Troll project — 27
- The Sean project — 32
- Project management and related considerations — 35
- Safety and the environment — 36
- Some differences between gas and oil projects — 36

4: GAS LAWS AND PROPERTIES — 39
- The nature of gases — 39
- The work of Robert Boyle — 40
- First steps in the liquefaction of gases — 40
- The behaviour of gases — 41
- Critical point (of temperature) — 43
- Van der Waals's equation — 44
- Joule-Thomson Effect — 44

	Liquefaction of methane by adiabatic expansion	45
	Liquefaction of methane by mechanical refrigeration	46
	Liquefaction of methane by the cascade process	46
5:	**GAS TREATMENT AND PROCESSING**	**47**
	Gas liquids and sulphur extraction: business scope	47
	Dehydration	47
	Hydrocarbon extraction	50
	Acid gases	54
	Typical flow sequence	55
	The Shell ADIP process	55
	The Shell Sulfinol process	56
	The Claus process	57
	The Shell Claus off-gas treating process (SCOT)	58
	Carbon dioxide removal	59
	Denitrogenation	60
	Mercury	61
	Helium	61
6:	**BASE-LOAD LIQUEFIED NATURAL GAS PLANTS**	**63**
	Operational base-load plants	63
	Plant components	65
	Gas receiving and metering	66
	Acid gas removal	66
	Mercury removal	67
	Dehydration	69
	Heavy hydrocarbon separation	69
	Fractionation	70
	NGL treatment	71
	Liquefaction	71
	Pure refrigerant cascade process	71
	Mixed refrigerant process	73
	Pre-cooled mixed refrigerant process	74
	Storage and loading facilities	76
	Utilities	77
	Safety	77
7:	**LNG AND OTHER TYPES OF STORAGE**	**79**
	LNG above-ground (and semi-buried) tanks	80
	LNG in-ground tanks	84
	Rollover	87
	Safety	88
	Seasonal storage	89
	Aquifers and depleted fields	91
	Salt caverns	93
	Rock caverns	95

8:	**LNG SHIP DESIGNS, OPERATION AND OWNERSHIP ASPECTS**	**97**
	Refrigerated ships: general aspects	97
	Design features: low temperature effects	98
	Types of cargo containment systems	99
	Metallic container materials	100
	Insulation materials	101
	Secondary barriers	101
	Ship construction regulations	102
	Self-supporting tank LNG designs	102
	Membrane tank LNG designs	105
	Internal insulation systems	107
	LNG ship designs	108
	Ship cargo operations	110
	Cargo transfer operations	111
	Safety	111
	Ship design and fleet configuration considerations	112
	Ship ownership and charter parties	113
9:	**LNG RECEIVING TERMINALS**	**115**
	Main components	115
	Site selection criteria	115
	Berthing and unloading facilities	117
	Types of storage	118
	Pumps	118
	Vaporisers	119
	Boil off gas facilities	120
	Metering and pressure regulation station	121
	Safety	121
	Gas quality	121
	Storage requirements	122
	Determination of seasonal stock	124
	Working capital	125
	Simulation models	125
10:	**TRANSMISSION SYSTEMS AND DISTRIBUTION GRIDS**	**127**
	Historical background	127
	System categories and characteristics	127
	Practical aspects of overland transmission pipelines	129
	Practical aspects of subsea pipelines	134
	Two phase and dense phase flow	135
	Compressor stations	138
	Basic pipeline economics	140
	Practical aspects of distribution grids	141
	Construction and commissioning	144
11:	**THE RESIDENTIAL MARKET**	**145**
	Evolution of the market	145
	Conversion	145

	Market characteristics	146
	Cooking	147
	Water heating	147
	Space heating	148
	Condensing boilers	151
	Heat pumps	152
12:	**THE COMMERCIAL AND TRANSPORTATION MARKETS**	**155**
	Market scope	155
	Horticulture	156
	Absorption type chiller/heaters	158
	Automotive uses	160
13:	**THE INDUSTRIAL MARKET**	**165**
	Radiant burners	165
	Direct reduced iron	167
	Oxygen enrichment	169
	Regenerative burners	169
	Fuel cells	170
	Lime kilns	172
	The emission challenge	172
14:	**POWER GENERATION**	**175**
	Combined cycle gas turbine	176
	CCGT cost and other aspects	178
	Repowering	180
	Combined heat and power	181
	Environmental considerations	183
	Relative benefits and gas values	185
15:	**GAS CONVERSION**	**187**
	Synthesis gas	187
	Steam reforming	187
	Partial oxidation	188
	Methanol	188
	Hydrogen	189
	Ammonia and urea	189
	Gasoline	191
	Middle distillates	192
	Methyl Tertiary Butyl Ether (MTBE)	195
	Other products	195
	Research and development	196
16:	**MARKET ASSESSMENT AND UTILISATION PRIORITIES**	**197**
	First priority: maximise local uses	198
	Second priority: exports as gas	199
	Third priority: conversion to other products	200
	A methodology for market assessment	201
	Assessing the large users' market	202

	The small industrial and commercial markets	204
	The residential market: general aspects	205
	Assessing a new residential market	205
17:	**FINANCIAL AND ECONOMIC ASPECTS OF PROJECT EVALUATION**	**207**
	Screening and ranking	207
	Measurements of profitability	208
	Cash flow components	208
	Payback (or payout)	209
	Accounting rate of return	210
	Discounted cash flow	210
	Net present value	210
	Internal rate of return (or earning power)	211
	Inflation	212
	Risk assessment	212
	Financing considerations	213
	Other considerations	214
18:	**MARKET STRUCTURES AND CONTRACTS**	**217**
	Pipeline gas market structures	217
	LNG market structures	219
	Contracts : some general considerations	221
	Main contract types : pipeline gas	222
	Contract type : LNG	224
	Contract provisions : pipeline gas	225
	Contract provisions : LNG	227
	Some concluding observations	228
19:	**NATURAL GAS PRICING**	**229**
	Some general concepts and principles	229
	Price components	231
	LNG pricing	234
	Pricing options	235
	The spot market and spot prices	238
	Summary	239

Natural Gas and other Energy Equivalents	240
Symbols and Abbreviations	242
Selected References	244

List of Figures

		Page No.
CHAPTER 1: WHAT IS NATURAL GAS?:		
1.1	Natural gas terminology	3
CHAPTER 2: EXPLORATION		
2.1	Hydrocarbon bearing sandstone rock	11
2.2	Typical anticlinal trap and spill point	12
2.3	Typical wedge-out stratigraphic trap	13
2.4	Three-dimensional seismic survey	14
2.5	Exploration drilling rig	15
2.6	Core sample	16
CHAPTER 3: PRODUCTION		
3.1	Terminology of reserves	22
3.2	Typical pre-development activities and inter-action	24
3.3	Launching the North Rankin 'A' jacket	25
3.4	North Rankin 'A' jacket in place	26
3.5	The completed North Rankin 'A' platform	27
3.6	Brent 'C' platform under construction	28
3.7	Brent 'D' Condeep gravity platform	29
3.8	Underwater manifold centre	30
3.9	Design concept for the Troll platform	31
3.10	Sean South jacket being installed	33
3.11	Sean South process and production platforms	34
CHAPTER 4: GAS LAWS AND PROPERTIES		
4.1	Gas manufacture	39
4.2	Gas standards	45
CHAPTER 5: GAS TREATMENT AND PROCESSING		
5.1	Gas drying unit, Den Helder, The Netherlands	48
5.2	St. Fergus gas plant, Scotland	51
5.3	Slug catcher, St. Fergus gas plant	52
5.4	Mossmorran fractionation plant	53
5.5	Shell ADIP process flow scheme	55
5.6	Shell Sulfinol process flow scheme	57
5.7	Shell SCOT process flow scheme	58
5.8	A SCOT plant	59
CHAPTER 6: BASE-LOAD LIQUEFIED NATURAL GAS PLANTS		
6.1	Schematic LNG plant: main components	66
6.2	North West Shelf LNG plant, Australia	67
6.3	Malaysia LNG plant: Bintulu, Sarawak	68
6.4	Two LNG trains under construction: North West Shelf project	70

6.5	Schematic pure refrigerant cascade process	72
6.6	Schematic mixed refrigerant process	73
6.7	Schematic pre-cooled mixed refrigerant process	74
6.8	Air coolers of one train: North West Shelf plant	75

CHAPTER 7: LNG AND OTHER TYPES OF STORAGE

7.1	Cross section of a typical above-ground LNG tank	79
7.2	LNG peak shaving plant at The Maasvlakte, The Netherlands	81
7.3	LNG storage, Bintulu, Malaysia	83
7.4	LNG storage, Withnell Bay, Australia	83
7.5	Cross section of a typical in-ground LNG tank	84
7.6	In-ground LNG tanks	85
7.7	In-ground storage	86
7.8	Schematic load duration curve	88
7.9	Schematic structure for underground storage	90
7.10	Underground storage at Bierwang, Germany	91
7.11	Wellhead at Bierwang	92
7.12	Making salt caverns for gas storage	93
7.13	Compressors for gas injection	94

CHAPTER 8: LNG SHIP DESIGNS, OPERATION AND OWNERSHIP ASPECTS

8.1	Principles of LNG ship construction	98
8.2	LNG ship containment systems: main design features	99
8.3	IHI-SPB free-standing tank system	103
8.4	Kvaerner-Moss free-standing spherical tank system	104
8.5	LNG ship construction, Nagasaki, Japan	104
8.6	'Northwest Sanderling' berthing at Sodegaura, Japan	105
8.7	Technigaz membrane tank system	106
8.8	Interior view of a Technigaz cargo tank	107
8.9	Gaz Transport membrane tank system	108
8.10	Two Gaz Transport membrane type ships	108

CHAPTER 9: LNG RECEIVING TERMINALS

9.1	Tokyo Gas's LNG terminal: Sodegaura	116
9.2	LNG terminal: simplified flow scheme	116
9.3	LNG unloading arms: Tokyo Gas's Sodegaura terminal	118
9.4	Open rack sea water vaporiser	119
9.5	Variation of LNG stocks in average year	123
9.6	Calculation of seasonal stock requirements	125

CHAPTER 10: TRANSMISSION SYSTEMS AND DISTRIBUTION GRIDS

10.1	A metering station Epe, Germany	129
10.2	A spread in action in Scotland	130
10.3	Coating and wrapping pipeline welds in The Netherlands	131
10.4	Laying a pipeline across a river	132

10.5	Lay barge 'ETPM 1601' in action offshore Western Australia	133
10.6	A reel laying ship in the North Sea	134
10.7	Submarine pipeline plough	135
10.8	A pipeline pig	136
10.9	Typical phase envelope for a natural gas mixture	137
10.10	Waidhaus compressor station, Germany	138
10.11	Interior view of the Ommen compressor station, The Netherlands	139
10.12	Effect of volume on costs	140
10.13	Effect of load factor on costs	140
10.14	Transportation cost for a particular diameter	141
10.15	Laying a large street mains at night	142

CHAPTER 11: THE RESIDENTIAL MARKET

11.1	Typical flueing arrangements	149
11.2	Schematic cross section of a condensing boiler	151
11.3	Hydro pulse condensing boiler & principle of pulse combustion	152
11.4	Heat pump concept	153

CHAPTER 12: THE COMMERCIAL AND TRANSPORTATION MARKETS

12.1	A gas turbine co-generation system	156
12.2	A gas engine heat pump system	157
12.3	Chiller/heater simplified flow chart	158
12.4	An absorption chiller/heater unit	159
12.5	A CNG refuelling station in Italy	162
12.6	Overnight CNG trickle feed refuelling	163

CHAPTER 13: THE INDUSTRIAL MARKET

13.1	Walking beam type reheat furnace	166
13.2	Direct reduced iron: Midrex process flow diagram	168
13.3	Scrap aluminium melting	170
13.4	Molten carbonate fuel cell power plant, San Ramon, California	171

CHAPTER 14: POWER GENERATION

14.1	World power generation fuel consumption 1990	176
14.2	Gas combined cycle electricity generation	177
14.3	Conventional coal-fired plant efficiency	178
14.4	Conventional fuel oil-fired plant efficiency	178
14.5	Conventional natural gas-fired plant efficiency	179
14.6	Natural gas combined cycle plant efficiency	179
14.7	Schematic gas-fired repowered combined cycle power plant: one unit	181
14.8	Typical cogeneration scheme	181
14.9	Co-generation plant: Tokyo	182
14.10	Average carbon dioxide emissions from thermal power plants	182
14.11	Average NO_x and SO_x emissions from thermal power plants	184
14.12	Comparative energy values and costs	184
14.13	A combined cycle plant at Waidhaus compressor station, Germany	185

CHAPTER 15: GAS CONVERSION
15.1	General view of an ammonia plant	189
15.2	ICI's ammonia plant, Severnside, Bristol, England	190
15.3	SMDS plant construction activities in Bintulu	192
15.4	Ethane cracker: Mossmorran, Scotland	194

CHAPTER 16: MARKET ASSESSMENT AND UTILISATION PRIORITIES
16.1	Principal activities involved in market assessment	204

CHAPTER 17: FINANCIAL AND ECONOMIC ASPECTS OF PROJECT EVALUATION
17.1	Illustrative project paybacks	209
17.2	An illustrative 'Star Diagram'	212

CHAPTER 18: MARKET STRUCTURES AND CONTRACTS
18.1	Classic pipeline gas market structure	217
18.2	USA: current marketing structure	219
18.3	Typical LNG base-load market structure	220

CHAPTER 19: NATURAL GAS PRICING
19.1	Value-cost-price	230
19.2	Influence of time-lag on wholesale prices	232

List of Tables

		Page No.
CHAPTER 1: WHAT IS NATURAL GAS?:		
1.1	Likely components of natural gas before processing	2
1.2	Natural gas compositions per cent by volume	4
1.3	Combustion characteristics	8
CHAPTER 2: EXPLORATION		
2.1	Geological time scale	9
CHAPTER 4: GAS LAWS AND PROPERTIES		
4.1	Examples of critical temperatures and pressures for selected substances.	43
CHAPTER 6: BASE-LOAD LIQUEFIED NATURAL GAS PLANTS		
6.1	Operational base-load LNG plants	64
CHAPTER 8: LNG SHIP DESIGNS, OPERATION AND OWNERSHIP ASPECTS		
8.1	Cryogenic materials and design cost factors	101
8.2	Percentage breakdown of LNG ship costs	109
8.3	Typical voyage losses percentage of ship's loaded cargo capacity	111
CHAPTER 9: LNG RECEIVING TERMINALS		
9.1	Preliminary calculation of base load terminal storage capacity	123
CHAPTER 12: THE COMMERCIAL AND TRANSPORTATION MARKETS		
12.1	Natural gas vehicle (NGV) statistics	160
CHAPTER 14: POWER GENERATION		
14.1	Typical average daily activity of a 2000 MW power plant	183
CHAPTER 15: GAS CONVERSION		
15.1	SMDS product variability	193
CHAPTER 17: FINANCIAL AND ECONOMIC ASPECTS OF PROJECT EVALUATION		
17.1	Illustrative summary of net present value calculation	211

PREFACE

The word 'gas' was invented by the Flemish scientist Jan Baptista van Helmont (1577-1644). In his work 'The Origins of Medicine' he presented the findings of his experiments (c 1609) during which he had discovered that a 'wild spirit' escaped from heated coal and wood. He stated: 'to this vapour, hitherto unknown, I give the new name "gas".' The derivation of the name is generally thought to be the Greek word 'chaos'. Apart from distinguishing gases from solids and liquids, van Helmont's further claim to fame was that he was the first to identify carbon dioxide as a separate substance.

However, the existence of natural gas, whatever it may then have been called, dates back over three thousand years when, it is believed, the Chinese collected gas coming to the surface naturally to heat pans of brine water to obtain salt. The Romans, Greeks and others in ancient times used seepages of gas to create ever-burning sacred lights for religious purposes. There are other references to natural gas in the literature through the Middle Ages to more recent times. But it was not until 1821 in Fredonia, New York, that natural gas was first used in a practical way in modern times.

During the latter part of the last century and through the early part of this one, the use of natural gas in the United States developed in parallel with the growth of gas manufactured from coal. It was not until 1935 that sales proceeds from natural gas overtook manufactured gas proceeds in that country. And it was not until the 1950s that natural gas finally displaced manufactured gas in North America.

Elsewhere, with a few exceptions, notably the Soviet Union and Canada where the use of natural gas started around the turn of this century, the natural gas industry is essentially a post World War II development. It has grown rapidly since then to the point where it now supplies some 20 per cent of the world's primary energy demand, occupying third position after oil and coal.

Many energy analysts predict a very bright future for natural gas with the expectation that its share of the world's primary energy demand will continue to increase over the coming decades. With an average annual rate of discovery some four times that of annual production on a global basis, it does not seem likely that the natural gas industry will be resource constrained over the foreseeable future, provided that it is economically, technically and politically feasible to develop this expanding resource base.

ABOUT THE BOOK

The main purpose of the book is to give the reader, who may be a newcomer to the natural gas industry or unfamiliar with certain of its commercial or technological aspects, a broad understanding of the industry's structure and activities from exploration through to the end-uses of natural gas. An alternative title could have been 'A Layman's Guide to the Gas Industry'.

It does not, nor can it, in the space available, attempt to describe in any great depth or detail every aspect and phase of the business; there is other published material available should the reader wish to delve more deeply into any particular subject. With this in mind, I trust that those readers who are experts in certain subjects will accept any instances where they may feel that the treatment given to their speciality is inadequate or incomplete. While I have received substantial assistance and helpful comments from many of my ex-colleagues in Shell, and have drawn heavily upon material published by various companies of the Royal Dutch/Shell Group of Companies and other entities, the responsibility for any inadvertent factual errors or misleading statements – and undoubtedly there will be some – is mine, not Shell's, and I apologise for them.

No attempt has been made to anticipate the development of possible new technological and commercial techniques, nor to forecast trends in supply, demand and related matters. This book is not about the future development and outlook of the industry, but about its basic nuts and bolts so to speak.

There are several practical difficulties in writing a book of this nature. These include, for example, what units of measurement, terminology, abbreviations, etc., one should use. Some of these are unique to the gas industry and as such may not always be readily comprehensible to those less familiar with the business. Moreover, despite the continued efforts of many national and international bodies, there is as yet no common practice within the industry itself on such matters. Accordingly, the author has used such units, expressions, terminology, abbreviations and the like which are either given in the reference literature or which seem most appropriate and understandable for the subject in question. To assist the reader, explanations are given where necessary and helpful.

Other difficulties have arisen since I started writing this book. Part way through the Soviet Union as we knew it disintegrated and is now in the process of being reconstituted, albeit in a different form, while changes were also taking place in East Europe, including the reunification of the previously separate states of East Germany and West Germany. For the purposes of this book I have been obliged to use that data available which in essence relate to the old Soviet Union (the USSR) and West Germany and they are referred to as such in the text.

Finally, the book has no index. This is deliberate and is because I considered that the contents list and lists of figures and tables were effectively an index for a book of this nature.

I was commissioned by Shell International Gas Limited to write this book shortly after I had retired from the Company. It has been a fascinating and challenging project to research and write. And despite its imperfections I hope that the reader will find some things in it which are new and of interest to him or her.

The Author March 1992

ABOUT THE AUTHOR

Malcolm Peebles was educated at Rutlish Grammar School, south London, where he matriculated. On leaving school in 1948 he joined the Anglo-Saxon Petroleum Company which was then and still is a wholly-owned subsidiary of the Royal Dutch/Shell Group of Companies. Later, after two years National Service when he received a commission in the Royal Artillery, he rejoined Shell and was initially concerned with crude oil scheduling, subsequently with aviation refuelling equipment.

From 1954 to the early 1960s he was concerned primarily with the inspection of existing markets and the surveying of new markets for liquefied petroleum gas mainly in Africa, the Middle East and South East Asia.

He was a founder member of Shell's natural gas organisation for areas outside North America when this was established in the early 1960s. Thereafter he was involved in almost all non-technical aspects of the pipeline gas, liquefied natural gas and natural gas liquids businesses until he retired from Shell in 1991. During that time he visited many gas markets throughout the world, lectured and presented some 90 papers on gas affairs to many national and international seminars, courses, institutions, associations and conferences. He has also written articles for various journals and four books including 'Evolution of the Gas Industry'.

Malcolm Peebles was appointed Director of Planning and Finance of Shell International Gas Limited in 1977 and a Director of Shell Coal International Limited in 1987.

On retirement from Shell he established his own consultancy company, Gas Advisory Services. He is a Companion of The Institution of Gas Engineers and a Fellow of the Institute of Petroleum.

ACKNOWLEDGEMENTS

I wish to express my gratitude to many friends and ex-colleagues in Shell whose assistance has helped me to make this book possible. I am particularly indebted to Roland Williams, Managing Director of Shell International Gas Limited, for asking me to take on the task and encouraging me throughout its compilation, also to Theo Oerlemans, Gordon Summers, Mike Brooks, Piet Zuideveld, Bob Herbert, Don Schafer, Roger Peres, Ralph Tutton, Colin Bowkley, David Strachan, Bert Pots, Harry McNicol, Pat Baldwin and others either for their constructive comments and/or for the trouble they have taken to seek out suitable reference material for me.

My special thanks are also due to Anne Biggs for her excellent and patient typing of my manuscript, to Tony Budd for reading my manuscript and pointing out where the text should be clarified to advantage and to the staff of Shell Publicity Services, especially Chris Andrews and Steve Rudd, who made many drawings and set out the text for printing.

In addition, a number of companies have kindly allowed me to use their photographic material to illustrate this book – my thanks to them.

Chapter 1

WHAT IS NATURAL GAS?

Natural gas is so called because it occurs naturally. It is formed from sediments rich in organic matter which in the past have been treated to very high temperatures and pressures. It is found in underground structures similar to those containing crude oil and may or may not be associated directly with corresponding accumulations of oil. Virtually all oil accumulations have natural gas associated with them, but many gas accumulations exist independently of any oil accumulation.

Associated and Non-associated Natural Gas
These are generic expressions which reflect the three main types of reservoirs or structures in which natural gas is found.

Gas from structures from which only gas can be produced economically is called NON-ASSOCIATED GAS (or unassociated gas).

Gas from condensate reservoirs which yield relatively large amounts of gas per barrel of light liquid hydrocarbons is also called NON-ASSOCIATED GAS. Although many condensate reservoirs are produced primarily for gas, there are cases where gas is re-injected or 're-cycled' to improve liquid recovery, particularly if no market for the gas is yet available.

Gas from reservoirs where it is dissolved in crude oil (SOLUTION GAS), and in some cases also in contact with underlying gas saturated crude (GAS-CAP GAS), is called ASSOCIATED GAS. Gas-cap gas is almost never produced until most of the economically recoverable oil has been extracted. Associated gas production rates will depend on the rate of oil production with oil usually representing the major part in energy equivalent terms.

Composition and Terminology
Natural gas usually consists largely of varying proportions of hydrocarbons of the paraffin series, with methane predominating.

If little or no pentanes and heavier fractions (C_5+) are present, the term 'dry gas' may be used. Conversely, 'wet gas' implies that significant quantities of C_5+ are contained in the gas as produced. Despite their use, indeed they are used in a few instances elsewhere in this book, 'wet' and 'dry' are not desirable expressions because of their ambiguity in that for gas marketers the term 'dry' usually denotes that the gas contains no water vapour.

Table 1.1 **Likely Components of Natural Gas Before Processing**

Name	Chemical Formula	Boiling Point at Atmospheric Pressure	
Methane	CH_4	−161.5°C	
Ethane	C_2H_6	−88.5°C	Gaseous at ordinary atmospheric temperature and pressure
Propane	C_3H_8	−42.2°C	
Isobutane	C_4H_{10}	−12.1°C	
Butane *(normal)*	C_4H_{10}	−0.5°C	
Isopentane	C_5H_{12}	27.9°C	
Pentane *(normal)*	C_5H_{12}	36.1°C	Liquid at ordinary atmospheric temperature and pressure
Hexane *(normal)*	C_6H_{14}	69.0°C	
Heptane *(normal)*	C_7H_{16}	98.4°C	
Octane *(normal)*	C_8H_{18}	125.6°C	

Other components such as carbon dioxide, nitrogen, hydrogen sulphide, helium, argon, mercury, water, solid particles, etc., may be present, as well as some very small amounts of other hydrocarbons with higher boiling points than octane.

The C_5 and heavier hydrocarbon fractions, which are liquids at atmospheric temperature and pressure, are commonly known as CONDENSATE or NATURAL GASOLINE, while ethane, propane and butanes, when extracted, are collectively known as NATURAL GAS LIQUIDS or NGL.

There are yet other terms. LIQUEFIED NATURAL GAS (LNG) is natural gas, mainly methane, which has been liquefied by cooling to minus 161.5°C at atmospheric pressure. LNG is not to be confused with LIQUEFIED PETROLEUM GAS (LPG) which is any mixture in either the liquid or gaseous state of propane and butanes.

Most of the above terms are illustrated in Figure 1.1.

Quality Considerations

Most natural gases produced contain water. This water has to be removed to prevent corrosion in pipes, appliances and other equipment. Moreover, water vapour may freeze in cold weather or form hydrates – a solid compound resulting from the combining of a hydrocarbon and water under certain conditions – which would obstruct the flow of gas. For much the same reasons, any entrained solid particles (sand, dust, etc.) are removed at the point of production.

The need to extract hydrocarbons, other than methane, and non-hydrocarbons from natural gas before it is transported and delivered to end-consumers will depend on the extent to which they are present in the gas as produced, the need to satisfy customer gas quality specifications, and any opportunities to commercialise (add value) by the separate sale of extracted components.

If substantial quantities of propane and heavier hydrocarbons are present these are most likely to be extracted, both as a commercial opportunity in its own right, and to avoid the risk of condensation in natural gas transmission and distribution systems. In some instances, the extraction and sale of these hydrocarbons can generate more revenue than the sale of the resultant 'dry' gas.

Figure 1.1 **Natural Gas Terminology**

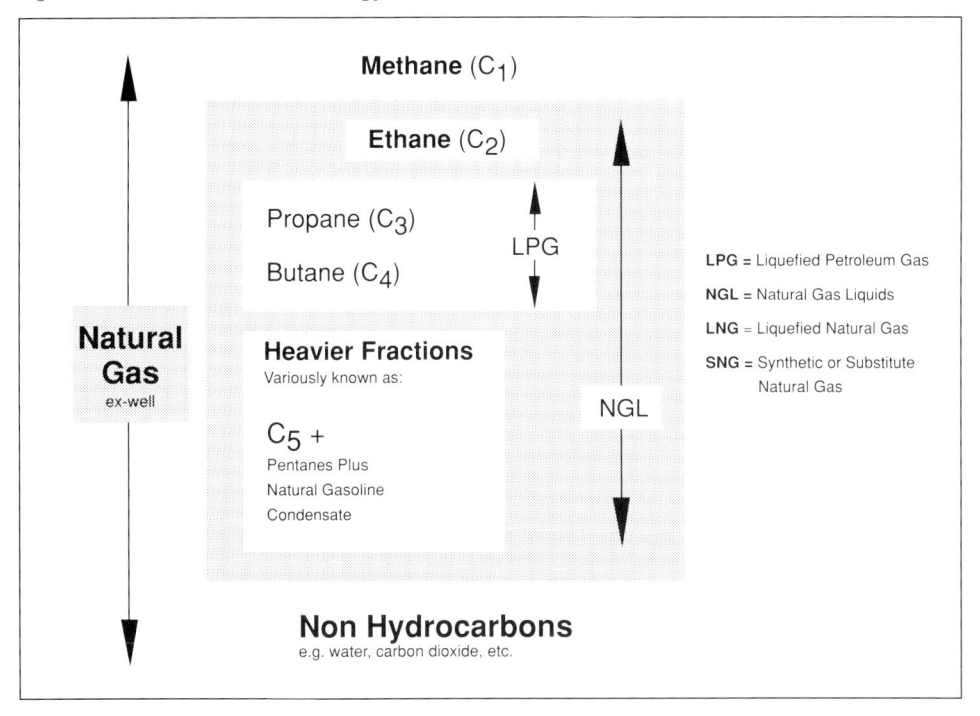

```
                            Methane (C₁)

                            Ethane (C₂)

                    Propane (C₃)
                                      LPG
                    Butane (C₄)               LPG = Liquefied Petroleum Gas
                                              NGL = Natural Gas Liquids
     Natural                                  LNG = Liquefied Natural Gas
     Gas        Heavier Fractions             SNG = Synthetic or Substitute
     ex-well   Variously known as:                  Natural Gas
                                      NGL
                  C₅ +
                  Pentanes Plus
                  Natural Gasoline
                  Condensate



                       Non Hydrocarbons
                       e.g. water, carbon dioxide, etc.
```

If the gas contains a significant amount of ethane, some of this component may be extracted for separate sale as a chemicals feedstock. The percentage of ethane extracted will be conditioned, inter alia, by the costs involved and the need to meet the calorific value of the gas to be distributed to end-consumers. The extraction of ethane and other gas liquids has been a major business activity in North America for many years and is growing rapidly elsewhere in the world.

Carbon dioxide must be kept below certain limits as it can cause corrosion in pipelines in combination with water, and excessive heat to be liberated in hydro-desulphurisation when natural gas is used as a chemicals feedstock. Many gas sales contracts stipulate that the carbon dioxide content cannot exceed 2 per cent by volume.

Hydrogen sulphide is highly corrosive and toxic. In many countries, the hydrogen sulphide content of gas for general sale is limited to a maximum of 5mg/Nm^3 (about 3 parts per million by volume). Sulphur extraction is a large, long established business in Canada where many of the gas deposits have a high hydrogen sulphide content. The Lacq gasfield in southern France is another example.

Nitrogen is neither corrosive nor toxic, but it has no calorific value. If, therefore, it is left in the gas it will reduce the calorific value of the gas and increase unit transportation costs because larger diameter pipelines will be needed to carry the same amount of useful energy. Whether to extract any nitrogen present or to leave it in the gas will depend on local circumstances. Interestingly, when the development of the huge Groningen non-associated gasfield in the Netherlands was being planned in the early 1960s, it was decided not to extract the 14 per cent nitrogen that that gas contains.

Although many natural gases contain very small amounts of helium, and in some instances argon, it is seldom economic (or necessary) to extract them other than in the United States, the main market for these gases.

Some natural gases contain traces of mercury and this must be removed if the gas is likely to come in contact with aluminium or aluminium alloys, for example in a cryogenic heat exchanger – see Chapter 6.

There are, of course, some exceptions to the foregoing where the cost of gas treatment/clean up cannot be justified and where the end-user, e.g. a large industrial consumer or a conventional gas-fired power plant, is able and willing to accept gas in virtually its original state. Otherwise, the only alternatives may be to leave the gas in the ground for possible future use, re-injection, which is costly and does not generate any revenue, or flaring (associated gas) if permitted by government.

Some examples of the differences in the composition of natural gas before processing, other than the extraction of water and solid particles, are given in Table 1.2. To underline the point, the examples quoted are all taken from producing gasfields in Europe. Similar differences exist in many other gas-bearing geographic areas.

Table 1.2 **Natural Gas Compositions per cent by Volume**

Constituents	Lacq France	Groningen Netherlands	UK North Sea (southern)	UK North Sea (northern)	Ekofisk Norway
Methane	69.52	81.30	95.05	78.8	85.2
Ethane	2.82	2.85	2.86	10.1	8.6
Propane	0.80	0.37	0.49	5.7	2.9
Butanes	0.60	0.14	0.17	2.2	0.9
Pentanes +	0.91	0.09	0.05	1.4	0.2
Hydrogen sulphide	15.59	trace	trace	trace	trace
Carbon dioxide	9.76	0.89	–	1.1	1.7
Nitrogen	–	14.35	1.26	0.7	0.5

The compositions indicated for UK North Sea fields are representative of a typical non-associated gasfield in the southern sector and an associated gasfield in the northern sector. Most associated gases contain higher proportions of ethane and heavier hydrocarbons than non-associated gases. The percentages given above do not add up to 100 per cent in all cases due to rounding off and/or traces of other constituents not stated.

Processes to extract various components from natural gas are described in Chapter 5.

Gas Treatment

Other than for chemicals feedstock and gas conversion purposes, natural gas after processing by the producer normally requires no further treatment before it is used. However, if a gas utility receives supplies of different qualities from several sources, it may need to treat some of these gases to make a harmonious interchangeable gas of marketable quality. This could, for example, include the injection or removal of nitrogen or the addition of propane or a propane-air mixture to lower or raise the calorific value as necessary. Filters are usually installed on customers' premises at the inlet to meters to remove any dust and solids picked up from the walls of pipes, but this can hardly be called treatment in the usual meaning of the word.

Combustion Characteristics

The CALORIFIC VALUE of natural gas defines how many heat units, e.g. kilocalories (kcal), British thermal units (Btu), joules (J), etc., will be released when a unit volume, e.g. cubic metre, cubic foot, etc., of gas is burned. Values may be quoted either gross or net. Briefly, 'gross' means that the water produced during combustion has been condensed to liquid and has released its latent heat, while 'net' signifies that such water stays as vapour.

Calorific value is important because the fuel requirement of an appliance is calculated by its designer as a heat input rate (e.g. kcal/h). However, he must convert this into an equivalent fuel volume to determine the sizes of pipework, valves, etc., required. To do this he uses the following expression:

$$\text{Volume (Nm}^3\text{/h)} = \frac{\text{heat input rate (kcal/h)}}{\text{calorific value (kcal/Nm}^3\text{)}}$$

A gas utility will use the same expression to convert its customers' consumptions into cubic metres per hour to calculate the required size of its distribution system.

Moreover, as a gas meter only measures the *volume* of gas delivered to a consumer, the calorific value must be specified so that the cost per heat unit can be calculated. For example, British Gas's quarterly residential customer accounts show the meter reading in cubic feet and the average calorific value of the gas supplied, typically about 1020 Btu/ft^3 (or say 38.1 MJ/m^3). In the UK, gas supplied is priced at so much per therm (i.e. 100,000 Btu). So applying the calorific value to the volumetric meter reading, dividing by 1000, and multiplying the result by the price per therm, gives the cost of the total amount of gas used.

In many countries legislation sets fairly narrow limits in permitted variations of the calorific value of gas supplied to end-users.

SPECIFIC GRAVITY. Natural gas (methane) is lighter than air, whereas propane and butane (LPG) are heavier than air. Thus in the event of a leakage from a pipe or an appliance, natural gas will tend to rise and disperse while LPG will fall to the ground and collect in a low point. These characteristics will govern the ventilation requirements and permitted locations for appliances using natural gas, LPG or other gases.

The WOBBE NUMBER (or Wobbe Index) of natural gas is defined as:

$$\frac{\text{Calorific Value}}{\sqrt{S}}$$

where 'S' is the specific gravity (air = 1). Wobbe Number is important for the following reasons.

In many appliances, the gas input rate is controlled by a metering orifice or 'jet'. If gas is supplied to the appliance at a fixed pressure, the flow in cubic metres per hour through any particular jet is proportional to:

$$\frac{1}{\sqrt{S}}$$

To the designer and user of the gas appliance, the important parameter is the heat input, not the volume of gas consumed, and this heat input equals:

$$\text{gross calorific value} \times \text{input in m}^3/\text{h}$$

Therefore, for a constant gas supply pressure and a given jet the heat input will be proportional to:

$$\frac{\text{gross calorific value}}{\sqrt{S}}$$

that is, heat input is proportional to the Wobbe Number.

If natural gas is supplied to an appliance designed for low calorific value manufactured gas, the heat input rate would be almost doubled, because the Wobbe Number of natural gas is nearly twice that of a typical manufactured gas. Wobbe Number is also important for calculating the amount of air to be drawn into simple burners such as those used in most domestic appliances.

The Wobbe Number should normally be maintained within about 5 per cent of its nominal value to ensure trouble-free operation of customers' appliances, although the exact value depends on the actual appliance. Instantaneous water heaters are usually sensitive to increases in Wobbe Number, because their combustion chambers are small, as also are some older appliances. However, a Wobbe Number variation from say +8 to −11 per cent of the design value may be acceptable for short periods, for instance when additional gas supplies are fed into the distribution system during periods of peak demand.

Because the Wobbe Number equals gross calorific value divided by $\sqrt{S}$, it is possible to produce a mixture of propane-air which has the same Wobbe Number as natural gas. Propane-air can be added to natural gas during peak demand, provided that the specific gravity of the resultant mixture is less than 1.0. This is to ensure that safety ventilation requirements are not compromised: there may be some well-ventilated, end-use locations where this is not a critical issue.

As propane-air has a much higher calorific value, depending on its air content, than natural gas, customers use fewer cubic metres for the same number of calories consumed. And as gas meters measure volume, customers will pay less money per kilocalorie of propane-air consumed. This reduction in the gas utility's proceeds is known as 'thermal giveaway'.

The use of propane-air introduces oxygen into the gas mains which can cause corrosion in the presence of water. Typical limits for oxygen are less than 0.5 per cent by volume when water vapour is present or below 3 per cent when it is not. Finally, the presence of oxygen can cause problems for some consumers, e.g. chemical plants, where changes in the composition of the gas supplied can be critical.

FLAMMABILITY LIMITS of natural gas in air are typically 5 to 15 per cent by volume.

AIR REQUIRED FOR COMBUSTION is approximately the same for all gases, provided that the air required is expressed in cubic metres per kilocalorie supplied.

Large differences in FLAME SPEED (burning velocity) between gases can result in faulty operation of the burner, such as flame lift-off and extinction or flash back to the primary jet, when switching from natural gas to other gases. But differences between various natural gases are small.

Examples of the above combustion characteristics are given in Table 1.3. Gas from the Leman field in the southern North Sea is more representative of many natural gases than, for example, Dutch Groningen gas with its high nitrogen content. However, the latter is included in the table as Groningen gas is a major component of mainland Europe's total gas supply portfolio. For comparison purposes, details are also given of a typical manufactured gas and of propane.

Table 1.3 **Combustion Characteristics**

Gas Property	Manufactured Gas	Groningen	Leman	Propane
Gross calorific value kcal/Nm^3	4770	8400	9770	23600
Relative density air = 1	0.5	0.643	0.587	1.52
Wobbe Number kcal/Nm^3	6750	10500	12800	19200
Max. flame speed metres/second	1.0	0.30	0.35	0.37
Combustion air required				
volumes	4.7	8.40	9.84	23.8
Nm^3/1000 kcal	0.98	1.00	1.01	1.01
Flammability limits % vol gas in air	5–35	6–17	5–15	2–10
Theoretical max. flame				
temperature in air: °C	1960	1920	1930	2000

Some of the matters covered in this chapter will be discussed in greater detail in subsequent chapters.

Chapter 2

EXPLORATION

Some Basic Geological Facts

The earth's crust basically consists of oceanic crust, which is relatively dense and thin (10 to 20 kilometres) and entirely below sea level, and continental crust which is relatively light and thick (25 to 50 kilometres). Oceanic crust is not known to be older than about 190 million years, i.e. the Jurassic period – see Table 2.1, whereas it is believed that rocks as old as 4500 million years can occur in the continental crust.

By studying the sequence of the rocks and the processes that deposited them, the geologist can, with the aid of fossil remains, usually assign a particular layer or formation to a period of geological time.

Table 2.1 **Geological Time Scale**

Era	Period	Epoch	Approx. age in millions of years
CENOZOIC (Cenos=recent) (Zoe=life)	QUATERNARY	Holocene Pleistocene	
	TERTIARY	Pliocene	7
		Miocene	25
		Oligocene	38
		Eocene	55
		Palaeocene	65
MESOZOIC (Mesos=middle)	CRETACEOUS	Maastrichtian Campanian Santonian Coniacian Turonian Cenomanian	100
		Albian Aptian Barremian Hauterivian Valanginian Berriasian	135

MESOZOIC (continued)	JURASSIC	Purbeckian Portlandian Kimmeridgian Oxfordian Callovian	163
		Bathonian Bajocian	173
		Toarcian Pliensbachian Sinemurian Hettangian	190
	TRIASSIC	Rhaetic Keuper Muschelkalk Bunter	230
	PERMIAN	Zechstein Rotliegendes	280
PALAEOZOIC (Palaios=old)	CARBONIFEROUS	Stephanian Westphalian Namurian	325
		Visean Tournaisian	345
	DEVONIAN		395
	SILURIAN ORDOVICIAN CAMBRIAN		440 500 570
PROTEROZOIC (Proteros= earlier)	PRE-CAMBRIAN		1000+

Note to Table 2.1: Not shown is the further sub-division of some Periods and Epochs into Upper, Middle and Lower.

Rocks are divided into three main groups:

i) *igneous rocks, e.g. granites and volcanic rocks;*

ii) *sedimentary rocks which are either fragments of other rocks deposited on land or under the sea by wind and water, chemically precipitated from evaporating waters, or are of organic origin; and,*

iii) *metamorphic rocks originally of igneous or sedimentary origin whose composition and structure have been profoundly changed by heat and pressure.*

Traces of animal and plant life are often preserved in sediments which have been deposited preferentially in topographic depressions known as sedimentary basins.

Natural gas and oil are derived from organic-rich source rocks that contain or have contained the remains of marine algae, bacteria and plant matter. Gas and oil mostly occur in the pore spaces of sedimentary rocks, although they can also be contained in fractured igneous and metamorphic rocks, and are trapped there if prevented from migrating elsewhere. Hydrocarbons are said to migrate when they leave the source rock in which they were generated and move upwards or sideways from areas of higher to lower pressure either to a reservoir rock where they are trapped or, if not, to the earth's surface where they escape.

Figure 2.1 **Hydrocarbon Bearing Sandstone Rock**

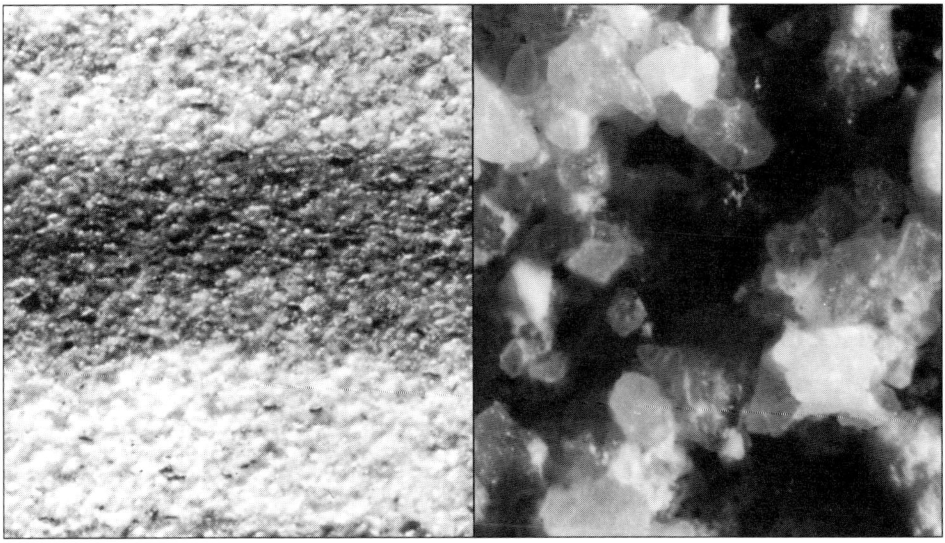

The photograph on the right is a ten-fold magnification of the pores in the sandstone on the left.

It is generally accepted that with bacterial action, together with increasing burial and hence temperature, the organic matter (kerogen) transforms at a given temperature initially into oil and at greater burial temperatures into gas. The type of organic matter determines whether it is capable of generating predominantly oil or gas. Hydrogen-rich, amorphous, sapropelic organic matter is an excellent source for both oil and gas, while hydrogen-poor, coaly organic matter, originating largely by the burial of forest and swampy types of vegetation, is mainly a source for gas.

Reservoir rocks must be porous and permeable if they are to store hydrocarbons and permit flow into wells. Porosity is usually expressed as a percentage of unit rock volume, with over 15 per cent generally considered to be good. Gas and oil will not occupy the entire pore space as water, usually saline, will adhere to at least 10 per cent of the rock grains. Permeability, which is expressed in units of Darcies, is a measure of the ease with which a fluid flows through the connected pore spaces of a reservoir.

Oil and Gas Accumulations

Oil and gas can accumulate underground if the following essential geological conditions are satisfied:

i) *The presence of reservoir rock, i.e. formations containing interconnected pores, e.g. sands and sandstones, or cracks and voids, e.g. some limestones.*

ii) *The presence, at the top of the reservoir, of a formation, i.e. a 'cap rock', that is impervious to the passage of hydrocarbons, e.g. clays, anhydrite, salt or shales.*

iii) *The presence of a trap, i.e. a geometrical configuration of the reservoir rocks and seal that prevents lateral escape.*

Because of a difference in density, oil will displace downwards the water previously filling the reservoir down to the spillpoint, and free gas (i.e. gas not dissolved in oil as a result of high pressure) if present, will collect in the highest part of the reservoir to form a 'gas cap' with the oil below it. Below the oil, the pores in the reservoir rock will remain full of formation water, usually saline. The same basic process applies if only gas is present.

Gas Traps

There are many types of traps which can be divided broadly into structural and stratigraphic traps. Structural traps result from the folding, faulting, or both, of the reservoir and cap rock. Figure 2.2 illustrates a typical asymmetric anticlinal trap in which the reservoir rock is capped by impervious rock. The latter also covers the

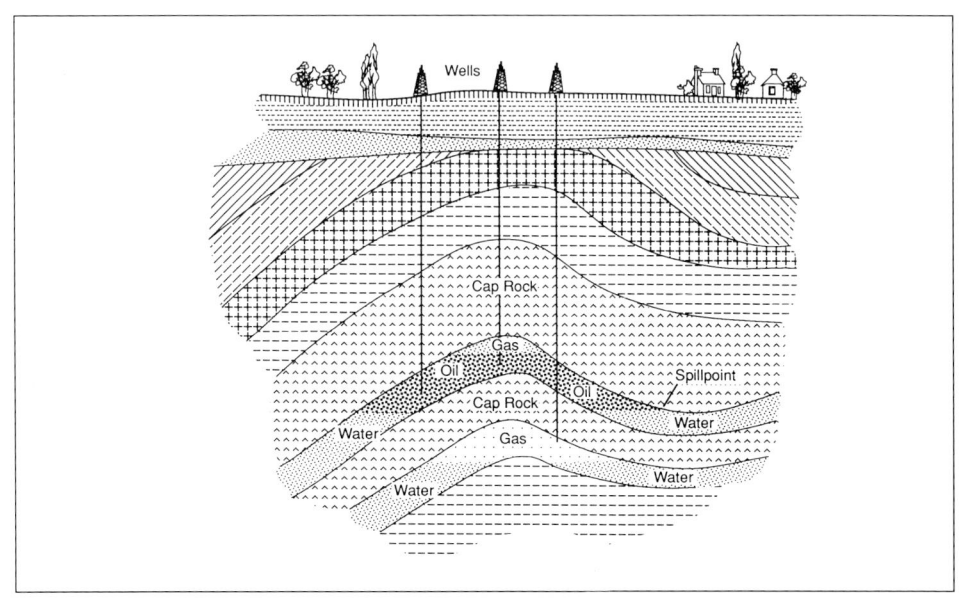

Figure 2.2 **Typical Anticlinal Trap and Spill Point**

flanks of the reservoir thus preventing the gas from escaping horizontally. Figure 7.9 (in Chapter 7) shows a trap in which the fault provides the closure for the tilted reservoir by bringing an impervious layer alongside it on the up-dip side. There are other ways in which structural traps can be formed.

Stratigraphic traps can be formed when reservoir rocks are pinched-out laterally or when they change to non-porous rocks. Such traps are often associated with the wedge-out of a sand layer in an up-dip direction and its replacement by impermeable clay or shale – Figure 2.3. Limestone is often impervious but may contain fissures and cavities that can form stratigraphic traps. There are yet other ways these traps can be created.

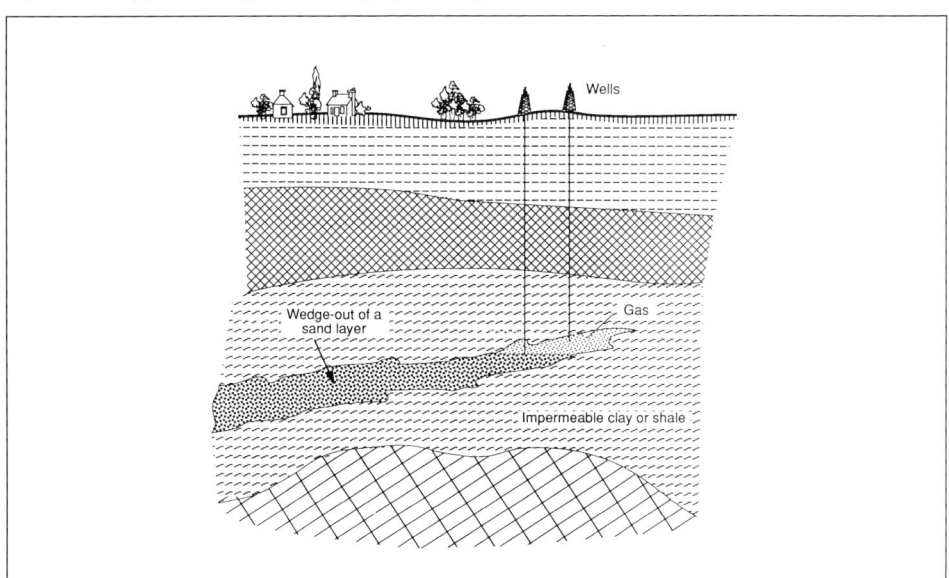

Figure 2.3 **Typical Wedge-out Stratigraphic Trap**

Exploration Methods

Field geology in modern exploration is used mainly to understand and predict the types of rock one may expect in the subsurface, in structures outlined by geophysical means. It is also customary to photograph the whole exploration (land) area from the air. By a stereoscopic study of these photographs, a reasonably accurate topographic map can be constructed. However, in many locations the deeper structure of the earth bears no resemblance to that seen on the surface.

The objective of geophysics is to determine the properties and structure of the rocks beneath the surface by the quantitative measurement of physical fields at the surface. The three principal methods used are gravity, magnetic and seismic.

The first of these involves gravimetric measurements of very slight variations in the force of gravity at the surface which is influenced in magnitude and direction by the distribution of rocks of different densities underlying the area. Gravity surveys can be carried out by air and are particularly useful for locating salt structures.

The magnetic method depends on measuring variations in the intensity of the earth's magnetic field. As sedimentary rocks are essentially non-magnetic, whereas igneous and metamorphic rocks are magnetic, this method can give an indication of the presence and thickness of a sedimentary formation. Airborne surveys make this a relatively inexpensive and quick method to map large areas.

Seismic

Seismic is the most important and expensive of the three. It works on the principle of low frequency sound waves generated from sources on or just below the land or water surface being reflected and refracted as they pass through rock layers, and measuring the time taken for the sound waves to travel back to the surface. Each layer bends the sound by a slightly different amount, depending on the difference in the speed of sound between one layer of rock and the next. Each rock interface, which has its own characteristic reflectivity, will reflect some of the sound waves back to the surface, while the refracted portion continues downwards to be reflected or refracted by deeper layers. These returning sound waves are picked up by geophones on the land surface, or by hydrophones in water, converted into electric signals and stored on magnetic tape for digital computer processing.

In new or little-explored areas, 2D (two-dimensional) seismic is generally used. Where detailed structural and reservoir information is required, or where the

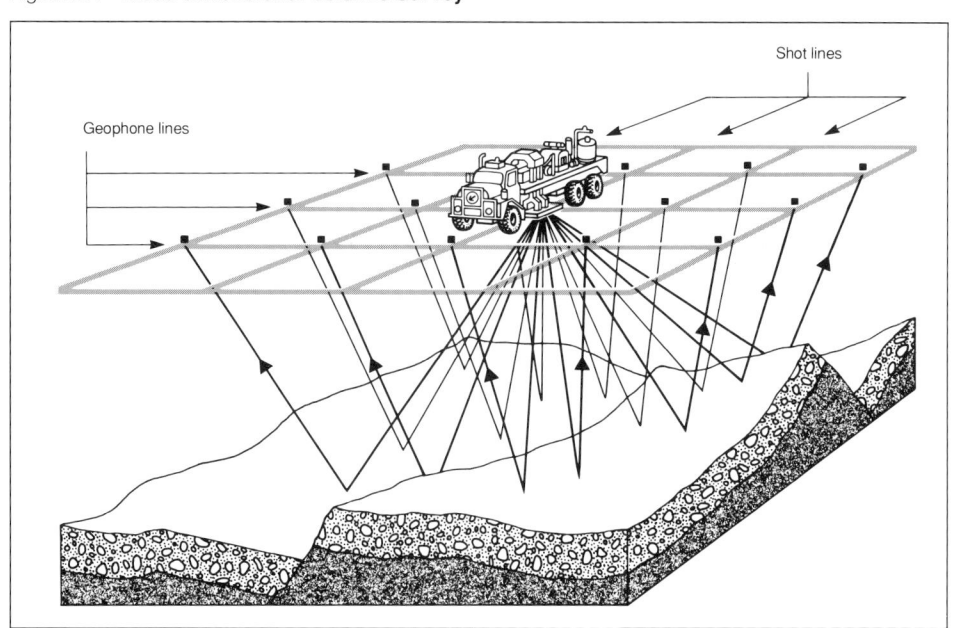

Figure 2.4 **Three-dimensional Seismic Survey**

geology is very complex, 3D (three-dimensional) seismic surveys are preferred. Shot over a close grid of lines 3D seismic is much more expensive, but it can reveal smaller, more subtle, hydrocarbon accumulations that might otherwise have been missed.

One of the important recent advances in seismic technology has been the ability of modern digital equipment to record many channels simultaneously – up to about 600 – to obtain greater subsurface coverage and resolution. Another is the availability of very powerful computers to handle the enormous number of seismic signals, some 30000 for each square kilometre of a 3D survey. Even so it is a complex and time-consuming process which can take several months. Computers are also used to assist the geophysicist in interpreting the results.

But despite the skills of the geophysicist, assisted by the latest technology available, the only way to confirm whether a structure does contain oil and/or gas is to drill a well. In addition to confirming the presence or otherwise of hydrocarbons, drilling provides more information on which to base further exploration and a future field development plan.

Exploration Drilling

Drilling techniques are the same, whether on land or at sea. Penetration of the subsurface is achieved by a steel, tungsten carbide or diamond bit attached to screw-jointed pipes, rotated at the surface. New lengths of drill pipe are screwed on as the bit descends. From time to time the drill pipe has to be withdrawn to replace

Figure 2.5 **Exploration Drilling Rig**

A 3000m well would typically require 130t of drill pipe, 200t of casing, 200t of drilling fluid, 150t of cement and 140t of fuel.

the bit or to take core samples. To prevent the hole caving in, a steel casing is inserted. The annular space between the casing and the rock is cemented to prevent fluid movement between strata. Pipe and casing diameters diminish as the well deepens.

Chemical mud is continuously pumped down the inside of the drill pipe and back up the annulus between the pipe and the side of the hole to the surface. This mud serves to lubricate and cool the bit, to flush out rock chips cut by the bit, to line the hole below the casing, and to prevent formation fluids flowing uncontrollably into the borehole and to the surface. Drilling mud and rock chips are continuously inspected for traces of hydrocarbons and core samples are taken occasionally for more detailed analysis.

Figure 2.6 **Core Sample**

Removing a core sample from the drill pipe during exploration operations in the North Sea.

After each section of the hole is completed and before it is cased, wireline log tools are used to plot certain electrical properties of the strata in contact with the well and to provide other data of a geological and geophysical nature, although they cannot detect directly the presence of oil and gas. However, the latter can be done with a drill stem test which stimulates the conditions of a completed well. Such tests can be carried out with or without the casing in place provided the casing, if in place, is first perforated by means of a gun lowered into the hole and fired electrically from the surface.

The first exploration well may reveal the presence of hydrocarbons that could prove to be commercial in due course. Equally, the first well may be 'dry' but it will still provide useful information of geological significance to justify further drilling. A number of dry wells may be drilled before exploration of the area in question is abandoned.

All wells drilled to discover hydrocarbons are called 'exploration wells', commonly known as 'wild cats'. A successful wild cat is a 'discovery well', an unsuccessful one a 'dry hole'. As a broad indication, and depending upon the nature of the terrain, location, etc., an onshore exploration well can cost up to $10 million, and an offshore well, particularly if it is in deep water, up to $20 million.

Appraisal Drilling and Commerciality Considerations

Once oil or gas has been discovered, additional wells must be drilled to establish the limits of the field; these are known as 'outstep' or 'appraisal' wells. As more appraisal wells are drilled, more data become available to confirm the nature and quality of the reservoir, e.g. its porosity, permeability, fluid saturations, gas or oil water contact, thickness, pressure, the extent to which wells may be in communication with each other, etc. From all these and other data, the geologist is then better able to define, with the possible help of further seismic surveys, whether the discovery is likely to be technically and commercially exploitable.

In the case of a natural gas discovery rather different considerations apply as to the assessment of its potential commerciality than to an oil discovery which is usually more straightforward to evaluate. For gas, the key questions to be resolved include: are there nearby market opportunities and, if so, what value (price) can the gas be expected to realise? What then is likely to be the cost of bringing the gas from the reservoir to its point of sale? Alternatively, if there is no nearby market, can one be developed and/or are there any export opportunities?

Such commercial assessments are normally outside the experience and knowledge of the explorer and appropriate marketing experts need to be called in at a relatively early stage to evaluate the nature and scope of the market opportunities for the (gas) discovery in question. This can be a lengthy, time-consuming process – see Chapter 16. In some instances, the outcome may be that although there are substantial gas reserves which are technically exploitable, there are no immediate economic commercially exploitable market opportunities. In that event, the decision may be taken to stop any further appraisal drilling and to log and cap those wells that have already been drilled, in effect to abandon that particular exploration effort until such time as a market opportunity develops. In extreme cases, if no market opportunity is likely to arise over the foreseeable future, the exploration effort, which was probably directed at discovering oil in the first instance, can be regarded to all intents and purposes as being abortive even though gas, but no oil, was discovered.

All exploration for hydrocarbons is a high risk activity with the commercial risks usually being greatest for gas, particularly in locations remote from market, for the reasons indicated above. An important consideration in any exploration venture which affects commercial viability if hydrocarbons are discovered, is the fiscal framework and related contractual arrangements within which the company has to operate.

Fiscal Framework

In virtually all countries mineral rights are vested in the state. Oil and gas industry operations can only be carried out under licences or contracts granted by governments. Important considerations for the exploring company are the investment at risk, i.e. the amount of work required and/or specified by government to explore the area adequately, the ability to manage operations, access to hydrocarbons if exploration is successful, and the economic return.

Licences are the traditional form of mining agreement where the government grants exclusive rights to a company, or a consortium of companies, to explore in a particular area. The company is responsible for financing the exploration and, if successful, is entitled to dispose of the production sometimes subject to deduction of royalty in kind. Government revenues come from this royalty and taxation.

An alternative arrangement which is becoming more common is a production sharing contract where the company takes the full exploration risk. It also pays all development and operating costs, which are recovered out of part of the production, referred to as 'cost gas (or oil)'. The remaining 'profit gas (or oil)' is then split in a predetermined manner between the company and the government. Royalty may also apply.

Risk contracts are where the company finances all exploration, in some cases all production costs as well, and is required to sell all production to the state oil and gas company and to take its return in cash or kind.

Association contracts involve a joint venture between the company and the state company, but otherwise are similar to a licence arrangement.

Basically governments obtain their revenues from a successful venture by various combinations of royalty as a percentage of production, royalty taken in kind, participating in the venture, corporation taxes or through special petroleum taxes. Windfall profits taxes are not uncommon in situations where realised prices for the hydrocarbons rise substantially above previously assumed expectations. Equally, there are examples of governments moderating their fiscal terms when they have been too onerous to attract exploration activity or when prices have fallen to such an extent that the venture is no longer economically viable.

Footnotes

Of necessity, in a book that endeavours to encompass the whole spectrum of the natural gas business, the foregoing is only a brief, simplistic and incomplete account of some geological aspects and activities involved in exploring for hydrocarbons. Exploration is arguably the most sophisticated and intellectually demanding task undertaken by the oil and gas industry. Knowledge of the earth's subsurface and exploration techniques are continually improving but, as stated earlier, exploration is a high risk venture; in the final analysis only expensive drilling can confirm if hydrocarbons are present.

On a worldwide basis, average annual rates of discovery of new gas reserves over the last decade or so have exceeded annual production rates by a factor of about four. On the one hand, this means that in global terms the life time, or reserves:production ratio, is increasing. On the other hand, as the majority of new gas discoveries in recent years have been made in areas remote from markets and thus are not always economic to develop, the need to find new gas reserves in exploitable locations remains a prime task. A further challenge for the gas industry is to find useful ways and means of utilising gas produced in association with oil in remote locations: it is estimated that the equivalent of more than 2 million barrels per day of associated gas is vented or flared.

To sum up, the world has a relative abundance of gas resources but many of these are in the 'wrong' place. The focus of gas exploration is to find more gas in the 'right' place.

Chapter 3

PRODUCTION

For the purposes of this chapter two assumptions have been made. The first is that the exploration venture was successful in that a gas discovery was made and that as a result of market studies the discovery is deemed to be commercial. However, in practice further seismic, appraisal drilling and the gathering of additional data will most probably be necessary to confirm this decision and, more particularly, to assist petroleum engineers, drilling and production specialists to delineate the reservoir more accurately and to draw up a preliminary pre-development plan.

For the sake of simplicity, the second assumption is that non-associated gas has been discovered, as defined in Chapter 1, although many of the activities and considerations described in this chapter would also apply to an associated gas discovery.

Later two actual projects are described for illustrative purposes.

Natural Gas Reserves

Before discussing what is involved in the production phase of a gas venture, it is appropriate to define what is meant by the term 'reserves'.

Unfortunately, there are no industry-wide generally accepted definitions of natural gas reserves and wide variations in both terms and principles are used by different companies and national and international bodies. For this reason, care should always be exercised when compiling reserves data taken from different sources and appropriate qualifications made. The practice within the Royal Dutch/Shell Group of Companies is indicated below and in Figure 3.1.

The EXPECTED ULTIMATE RECOVERY is the total volume of gas which may be expected to be recovered commercially from a given area at current or anticipated prices and costs, under existing or anticipated regulatory practices and with known methods and equipment. This total can be expressed as the sum of the 'ultimate recovery from existing fields' and the 'expectation from future discoveries'.

The ULTIMATE RECOVERY FROM EXISTING FIELDS is the sum of the 'cumulative production' from such fields and the 'remaining reserves from existing fields'.

The REMAINING RESERVES FROM EXISTING FIELDS is the sum of 'proven reserves', 'discounted (i.e. 50 per cent of) probable reserves' and 'discounted (i.e. 25 per cent of) possible reserves'.

PROVEN RESERVES represent the quantities of natural gas (and natural gas liquids if present) which geological and engineering data demonstrate with reasonable certainty to be recoverable in the future from known gas reservoirs. They represent strictly technical judgements, and are not knowingly influenced by attitudes of conservatism or optimism.

DISCOUNTED PROBABLE RESERVES are those quantities of natural gas (and gas liquids) for which there exists a 50 per cent probability that they will materialise. Such reserves are usually allocated to some conjectural part of a field or reservoir as yet undrilled or incompletely evaluated where, while characteristics of the productive zone and fluid content are reasonably favourable, other uncertain features may have equal chances of being favourable or unfavourable for commercial production.

DISCOUNTED POSSIBLE RESERVES are those quantities of natural gas (and gas liquids) thought to be potentially producible but where the chance of their being realised is thought to be of the order of 25 per cent. Reserves in this category are usually allocated to possible extensions of existing fields where, although geological data may be favourable, uncertainties as to the characteristics of the productive zone and/or fluid content are such to preclude a probability greater than 25 per cent.

Figure 3.1 **Terminology of Reserves**

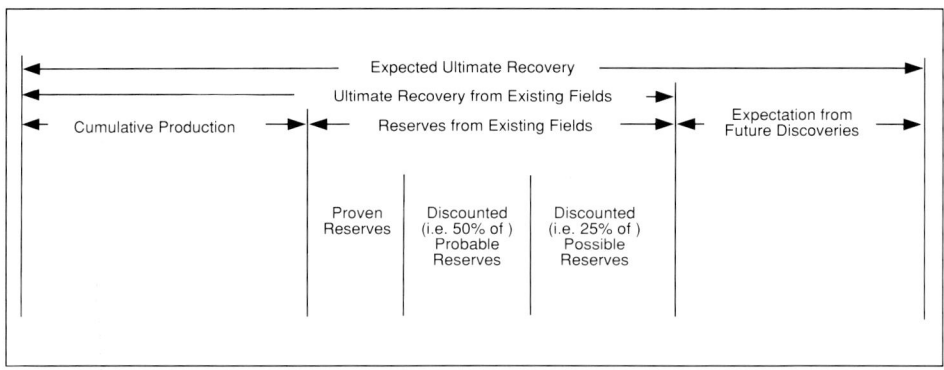

The term 'gas-initially-in-place' represents estimates of the total quantity of gas initially present in the reservoir(s), which in conjunction with various engineering and economic criteria, provide the basis for the 'reserve' estimate. It should not be confused with 'reserves' which word implies that physical recovery is possible.

In Shell, the above terminology and definitions also apply to crude oil reserves.

Pre-development Work
In many respects, the exploration phase of a gas venture is only the end of the beginning and not the beginning of the end – much remains to be done before gas can be produced and brought to market.

In the pre-development phase of production, work starts with the petrophysical assessment of reservoir rock properties e.g. porosity, permeability, etc., and fluid content obtained from measurements made in the discovery and appraisal wells. Potential reservoir intervals are selected for testing the production of gas; flow rates and reservoir pressures are measured and samples collected for laboratory analysis. These data are the particular concern of the reservoir engineer and the production engineer.

To state what may seem obvious, gas will only flow naturally from a reservoir when the pressure in the gas contained in the pores of the reservoir rock exceeds the pressure in the borehole at the bottom of the well. The rate of flow is determined by the differences in the pressure, the area of rock exposed to the borehole, the permeability of the rock and the viscosity of the gas. A productive reservoir consists of high permeability rock and plenty of natural energy to drive the gas to the surface.

Meanwhile, the reservoir geologist collaborates with the exploration geologists and seismic interpreters to construct a reservoir geological model which defines the accumulation in a quantitative manner. Further seismic will most probably be undertaken and more appraisal wells drilled to refine the structural interpretation and evaluate conditions over a wide area of the field, particularly if there are variations in the rock properties. Onshore appraisal wells can be used later for production, but those offshore are usually abandoned after testing.

As more data are gathered, the reservoir engineer predicts the likely performance of the reservoir in terms of flow rates over a long period. Meanwhile, the production engineer will be defining casing designs and well completion methods, selecting the appropriate downhole production equipment, ascertaining whether acid treatment or hydraulic fracturing of the rock is likely to be necessary, how to minimise reservoir damage caused by drilling, etc. He will also assess the theoretical effects of applying artificial methods, e.g. compression facilities, to improve recovery with the aim of maintaining a steady level of gas deliverability over the envisaged contract life.

Once the subsurface work and related studies are well advanced, the reserves have been estimated, the number of wells and their pattern established and a suitable production profile agreed, the provisional designs for the surface production facilities can be made. In this regard, gas quality is an important consideration as in addition to these production facilities, it may also be necessary to provide costly gas treatment/processing facilities in order to extract valuable liquid components and/or unwanted contaminants so as to produce a gas of marketable quality – see also Chapter 5.

Alternatively, if the gas discovery in question is rich in entrained gas liquids (condensate), but there is no market readily available for the gas, the development plan may be to re-cycle the rich gas to extract condensate at the surface and re-inject the processed lean gas into the reservoir. This has the advantage of storing temporarily unsaleable gas and maintaining reservoir pressure. The latter is desirable to prevent liquids forming and separating from the gas in the reservoir, an occurrence known as retrograde condensation, where subsequent recovery of the liquids under most reservoir conditions is not possible.

In practice, all this pre-development work involves considerable repetition and overlapping and a very close integration of many different skills and expertise as illustrated in Figure 3.2.

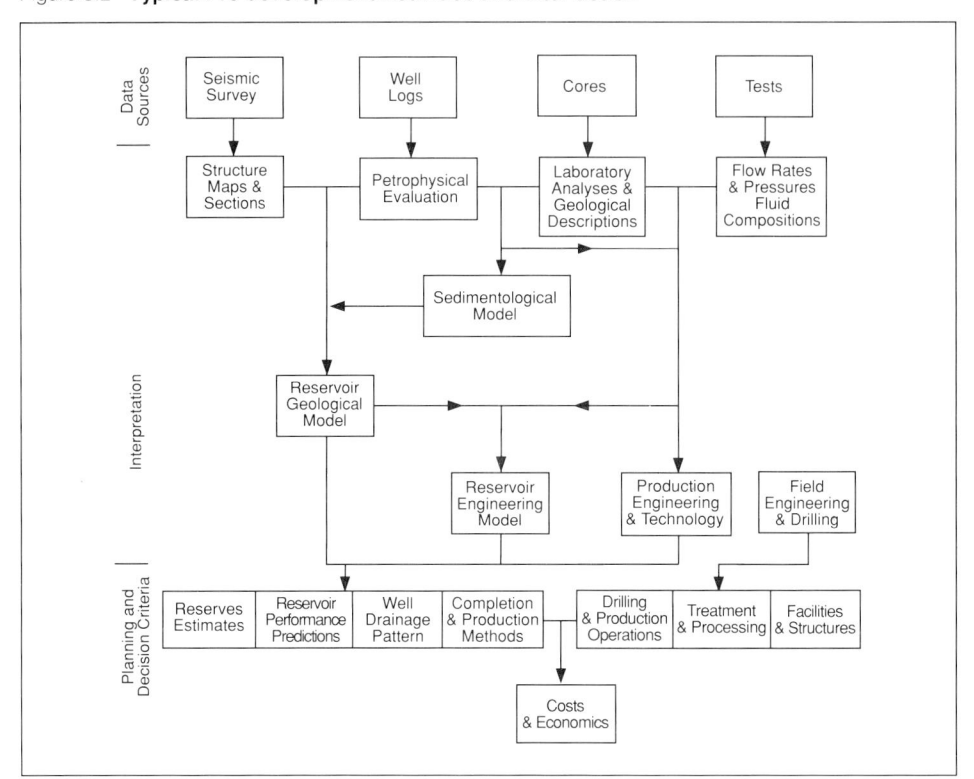

Figure 3.2 **Typical Pre-development Activities and Inter-action**

```
Data Sources:
  Seismic Survey → Structure Maps & Sections
  Well Logs → Petrophysical Evaluation
  Cores → Laboratory Analyses & Geological Descriptions
  Tests → Flow Rates & Pressures, Fluid Compositions

Interpretation:
  → Sedimentological Model
  → Reservoir Geological Model
  → Reservoir Engineering Model
  → Production Engineering & Technology
  → Field Engineering & Drilling

Planning and Decision Criteria:
  Reserves Estimates | Reservoir Performance Predictions | Well Drainage Pattern | Completion & Production Methods | Drilling & Production Operations | Treatment & Processing | Facilities & Structures
  → Costs & Economics
```

The Development Phase

Once the development of the project has been approved, all or at least parts of the pre-development work will be repeated in greater detail to provide an actual development plan for implemetation. The project then moves into the design, construction and installation phase with much of the work devolving on the facilities engineers.

If the gasfield is offshore, decisions will have to be taken on the type of platform to be installed, the number of producing wells to be drilled, whether any necessary treatment of the gas should be done on the platform or in an onshore plant, the route of the pipeline to landfall and many other similar matters of a practical/design nature. As this work proceeds, initial design concepts may be discarded in favour of more economic alternatives as will be illustrated later.

Platforms merit separate mention as they are the key component of an offshore production venture.

Offshore Production Platforms

The majority of offshore production platforms are steel-piled structures consisting of a superstructure, a jacket or tower and a foundation. Structures can range from a single unattended deck to a multi-level fully integrated structure with drilling equipment, process facilities, living quarters, a helicopter landing deck, etc.

Jackets are welded steel space frames which support the superstructure and provide a template for pile-driving. For deep water they may be built in several sections and installed sequentially one on top of the other. Towers, which comprise a small number of large diameter steel legs, perform the same function.

Jackets are usually transported to their location on a barge and either lifted off with a crane or launched directly into the sea, while towers can be floated to their location using the buoyancy of their large legs. Both are up-ended by controlled flooding of the legs. Normally, jackets and towers are pinned to the seabed by hollow steel piles – the foundation. The superstructure, or topsides as it is often called, is lifted on in modules once the jacket is in place and fully secured to the seabed.

Figure 3.3 shows the launching of the North Rankin 'A' jacket, Figure 3.4 with the jacket standing upright on the seabed, and Figure 3.5 the completed platform with its massive superstructure weighing 54000 tonnes and flare support structure which

Figure 3.3 **Launching the North Rankin 'A' Jacket**

Courtesy of Woodside Offshore Petroleum Pty. Ltd.

weighs 3130 tonnes. The platform stands in 125 metres of water and is 90 metres above sea level.

Concrete gravity structures were originally developed for the North Sea at locations where steel-piled structures would be extremely costly and difficult to install because of the nature of the environment and the limited 'weather window' for installation, e.g. pile-driving, etc. They sit on the seabed by virtue of their own weight. As all the vertical and horizontal loads are transmitted to the top layers of the seabed, this has to be well consolidated if a concrete structure is used.

Concrete structures of which there are many designs, have the merit of having most of their topside facilities already in place when they are towed out to their location. This reduces the time and cost of commissioning the platform. For oil production, concrete gravity platforms can have the further advantage of being able to store oil in the large cells which may form the lower part of the structure.

Figure 3.4 **North Rankin 'A' Jacket In Place**

Courtesy of Woodside Offshore Petroleum Pty. Ltd.

Figure 3.5 **The Completed North Rankin 'A' Platform**

Courtesy of Woodside Offshore Petroleum Pty. Ltd.

Other types of platform include steel gravity structures, buoyant towers, guyed towers, tension leg platforms and floating platforms. To some extent these are self-explanatory from their names; they are not discussed further here, nor are the various types of underwater manifolds (see Figure 3.8) that are used where conventional platforms would be uneconomic/impracticable and for developing parts of a field beyond the reach of existing platforms.

As it is impracticable in a book of this nature to cover all the very many different activities involved in the production phase of a gas venture, it was felt that brief descriptions of two actual projects would help to give the reader an overall impression of what happens in practice. The two selected are both offshore projects as these are generally more complex, at least in terms of design and technological aspects, than onland projects.

The Troll Project

Troll was discovered in 1979 by A/S Norske Shell as operator for a joint venture in block 31/2 of the Norwegian sector of the North Sea. Natural gas was found in reservoir sandstone deposited in the middle to upper Jurassic. Reservoir rocks are

Figure 3.6 **Brent 'C' Platform Under Construction**

Built by Sir Robert McAlpine & Sons in Scotland. After completion it weighed 320000 tonnes with an overall height of 290m. The 36 storage cells are capable of holding 660000 barrels of oil. Large quantities of associated gas are produced from the Brent field.

Figure 3.7 **Brent 'D' Condeep Gravity Platform**

Under tow from Stavanger, Norway, to its North Sea location. Even with a towing draught of 80m the structure stands 130m above the sea. It weighs 210000 tonnes and has 48 well slots.

Figure 3.8 **Underwater Manifold Centre**

Built for Shell/Esso by Hollandse Constructie Groep at Schiedam, near Rotterdam, for installing in a North Sea field.

up to 400 metres thick and lie about 1400 metres below the seabed. The field was assessed by means of 26 exploration and appraisal wells and was found to extend into neighbouring blocks 31/3, 31/5 and 31/6. As a result of this exploration and appraisal work, reserves are estimated to be some 1200 Bcm, with some underlying oil, which makes Troll the largest offshore gasfield discovered to date in Europe.

During the mid 1980s, Statoil, on behalf of the co-venture partners of the field, negotiated sales contracts with Ruhrgas, Thyssengas and BEB of Germany, Gasunie of The Netherlands, Distrigaz of Belgium, Gaz de France, Ferngas of Austria and Enagas of Spain for a total maximum volume of approximately 650 Bcm with a maximum annual volume, including option quantities, of 36 Bcm. First deliveries would come from much smaller Norwegian fields, in particular Sleipner, commencing 1993, followed and eventually replaced by deliveries from Troll starting in late 1996.

While these commercial activities were taking place, field development design work continued. The initial concept envisaged a concrete structure platform on which would be placed gas treatment facilities. Subsequently, it was decided it would

Figure 3.9 **Design Concept for the Troll Platform**

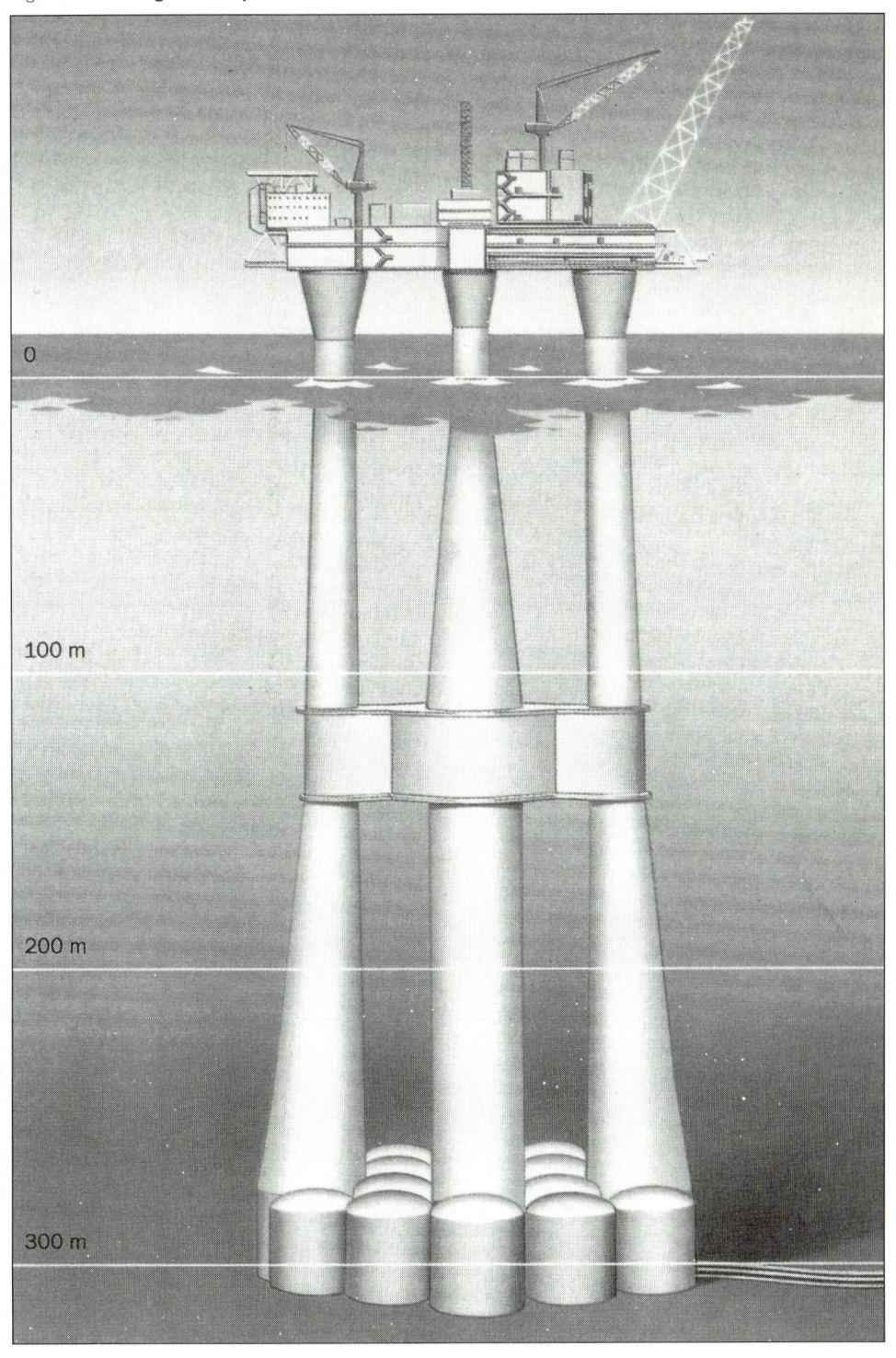

be more desirable to reduce the offshore manning levels thus necessitating a reduction in the offshore gas treatment facilities. This in turn meant piping the untreated gas some 60 km to a land-based processing plant to be built at Kollsnes, northwest of Bergen. Piping untreated gas over long distances is feasible thanks to recent advances in dense phase flow technology which is described in Chapter 10. One consequence of this decision is that the weight of the superstructure can be reduced from 40000 to 20000 tonnes.

The Kollsnes plant is designed to handle about 80 million m^3/day and will dry the gas and remove about 2000 tonnes of condensate per day before the gas is recompressed and transported by offshore pipelines to Emden, Germany, and Zeebrugge, Belgium, via an existing platform in the Heimdal field, the new platform being built for Sleipner and a new riser platform to be built near Sleipner.

The concrete platform – see Figure 3.9 – will be 430 metres high and stand in a water depth of over 300 metres. In this part of the North Sea wave heights can reach 30 metres. Contrary to the general comment made earlier, the seabed where the Troll platform will be installed is quite soft. To overcome this problem, the skirts on the concrete base will be pressed into the seabed to form a good foundation. The platform will have 39 production wells and one observation well drilled within a 500 metre diameter area immediately below the platform deviating out at angles into the reservoir from below the seabed. Each well will be capable of producing up to 3.4 million m^3/day although this will be limited to 2.8 million m^3/day (equivalent to a rate of about 1 Bcm p.a.) during normal production.

Total investment for the platform, drilling, pipelines to shore and the onshore plant up to 1997 is estimated to be about Norwegian Kroner 22 billion (in 1989 money).

In advance of the developments described above, Troll commenced supplying gas in 1991 through an underwater manifold and via a 50km pipeline to the Oseberg field to boost oil production. The total quantity to be supplied by this means is 25 Bcm over 11 years.

The Sean Project

Sean North and Sean South were discovered in 1969 in licence area P.054 in the southern part of the UK sector of the North Sea by Shell Expro as the operator for the then co-venture partners, Shell, Esso, Allied Chemicals and the National Coal Board (NCB). Since then Union Texas has taken over Allied Chemical's share and Britoil NCB's share. The fields are about 5 km apart, some 100 km off the Norfolk coast and 15 km southeast of the Indefatigable non-associated gasfield.

Each field contains about 6 Bcm of reserves, at a recovery factor of 80 per cent, in the Rotliegendes aeolian dune sands of the Permian basin at a depth of some 2500 metres below the seabed with reservoir thicknesses of about 75 metres.

Figure.3.10 **Sean South Jacket being installed**

On the deck of the Heerema derrick barge 'Balder' is the jacket for Sean North.

It was not until the early 1980s that market and economic conditions justified committing these reserves for sale to British Gas. Almost uniquely these two fields were contracted and developed for peak shaving purposes only, i.e. they would only produce gas, up to 17 million m^3/day, during limited periods of high demand in the winter months. Pricing and other contractual provisions reflected the rather exceptional type of service the Sean fields are required to provide.

The production development plan called for two 6-leg steel production platforms, one in each field, and one 8-leg steel process and accommodation platform in Sean South linked by a bridge to the production platform in that field – see Figures 3.10

and 3.11. Each production platform is designed to carry up to 12 producing wells, they weigh about 4500 tonnes each and stand in a water depth of some 30 metres.

The process platform has utilities, facilities for dehydration and CO_2 removal and living accommodation for 50 persons. It is linked to the production platforms with inter-field pipelines and to the land-based gas terminal at Bacton by a 30 inch pipeline. The platform weighs about 8800 tonnes with a deck clearance above sea level of 15 metres.

Figure 3.11 **Sean South Process and Production Platforms**

Courtesy of Shell UK Exploration & Production

In September 1982, Shell Expro established a project team which at its peak comprised 380 people. By August 1983, the conceptual design was completed, and with government approval being obtained in April 1984, jacket fabrication commenced followed by topsides fabrication in August. In December 1984, the offshore detailed design was finalised and two of the jackets were completed and installed. Drilling and pipeline laying commenced in February 1985. By the end of 1985 the pipeline to shore was completed, topsides installed and offshore hook-up started which was completed in July 1986. In August 1986, a maximum delivery capacity trial was carried out successfully. Hundreds of contractors were involved with over 1000 people employed just in pipelaying. The total project cost approximately £350 million.

Project Management and Related Considerations

The Troll and Sean projects described above are very unlike in a great many respects and neither of them can be regarded as being representative of a typical gas production project. Indeed it would be difficult to select any particular project as being typical as each has, as one would expect, its own unique geological, technological, topographical and economic features. But all such projects have the common need for careful integrated planning and co-ordination involving the use of sophisticated control and monitoring techniques, critical path analysis, materials procurement and contracting systems and similar management tools, many of which utilise computer programs.

A further complexity with gas projects is the essential need to co-ordinate the development and timing of the production phase with those parties responsible for developing other phases of an integrated gas project. The late completion of any part of a gas chain extending from the bottom of the well to the burner tip will have an adverse effect on the economics of the whole gas supply system. In this respect, no one part of an integrated gas system is any less or more important than any other. This, of course, also applies to the physical/technical integrity of every part of the total supply system.

The production phase of a gas project must also be certain that it can live up to its contractual obligations as regards the volumes and quality of gas to be supplied over the life of the contract which can be 20 years or more, and can meet any daily, weekly, monthly and/or yearly offtake rates that may be agreed with the gas buyer.

These requirements have an effect on the pre-development study work. The confidence level of the recoverable gas reserves estimate has to be very high and provision made in the field development plan for drilling more wells, introducing compression facilities, etc., at appropriate times in the life of the contract. For example, a 20 year 5 million tonnes p.a. LNG contract could require total reserves of at least 200 Bcm, after allowing for gas used for energy purposes in production, liquefaction and transportation. Moreover, the reserves must be capable of being produced at a fairly even and constant rate over the contract period. Likewise, for a pipeline gas contract the reserves required would not only depend upon the total contract volume to be supplied, but also the additional amount of gas needed for own use purposes, e.g. utilities, compressors, processing, etc., with the further complication of a much greater fluctuation in daily offtake rates than is normally the case for an LNG or gas conversion project.

Finally, project management is not confined to the foregoing considerations as due regard has to be taken of the views of any other co-venture partners. Many gas production ventures involve several entities who were joint concession holders before exploration commenced or who subsequently 'farmed-in'. While one entity is normally appointed to be the operator, other partners will need to be assured that the project is properly designed, executed and managed as they are required to pay their shares of the investment involved and accept the subsequent project risks.

Safety and the Environment
As with all gas activities safety and environmental conservation will have high priority from the outset so as to eliminate, as far as is technically and economically practicable, any possible accidents to operating personnel and the general public arising from equipment failures and the like, as well as any possible damage to the environment, whether onshore or offshore. The early identification of potential hazards can lead to their elimination, while a quantitative reliability analysis of equipment and materials can establish likelihoods of failure and enable the design engineer to select alternative equipment or a different design.

Environmental concerns will include the safe disposal of spent drilling materials and equipment, the visual impact and noise created by drilling activities in populated areas, careful siting of access roads and pipeline routes, etc. In many countries, environmental impact statements have to be filed with the local authorities and permits obtained before drilling and other work can commence. In some situations this can be a time-consuming process and may involve public hearings before permission is granted to proceed, and if not granted, alternative plans may then have to be prepared. All this has to be allowed for in the overall time schedule and critical path analysis.

Some Differences Between Gas and Oil Projects
The production of gas has many things in common with the production of oil but also some important differences. With non-associated gas, recovery rates can exceed 80 per cent of the gas-in-place or more than double that usually achieved for many medium to heavy crude oil reservoirs after applying secondary recovery techniques. For gas, production facilities frequently have to be designed to handle very high pressures of 1000 bar or more, although some oil reservoirs can also contain gas at high pressure. Oil fields can be developed step-wise over a period of years with production at any time being increased or decreased in line with demand trends. For gas, production facilities have to be designed and installed with the required infrastructure from the outset to meet contractual offtake rates and quality obligations many years ahead, and there has to be a high degree of confidence that sufficient reserves can be recovered to meet these obligations over contract life.

Gas production is but one part of a 'closed-loop' chain of supply, whereas oil supply is infinitely more flexible. If oil supplies from any one source are not available for any reason, there are many other sources that can be drawn upon quickly which is seldom the case with gas supplies. These and other differences all have an impact on the design and operating requirements of a gas production venture.

Chapter 4

GAS LAWS AND PROPERTIES

This chapter is of a historical nature; it is a resume of some selected basic gas laws and gas properties and the scientists who discovered them. Its purpose is to introduce and to provide a better understanding of subsequent chapters on gas processing and liquefaction.

The Nature of Gases
Matter may exist in three states – gas, liquid and solid. As the English physicist, Sir Oliver Lodge (1851-1940), so aptly put it: "A solid has volume and shape, a liquid has volume but no shape, and a gas has neither shape nor volume".

By definition, gases do not have a free surface. They fill any vessel in which they are placed, even if this means a change of pressure within the vessel in which the gas is contained. In the simplest possible terms, the nature of gases can only be explained as being composed of extremely minute particles which are far apart and in constant, chaotic motion. Whereas gases are highly compressible, liquids are relatively incompressible and do not necessarily fill any vessel in which they are placed as they retain their own volumes. But both gases and liquids are fluids in that they lack the rigidity of the solid state. Gases can ordinarily be formed into liquids by either cooling

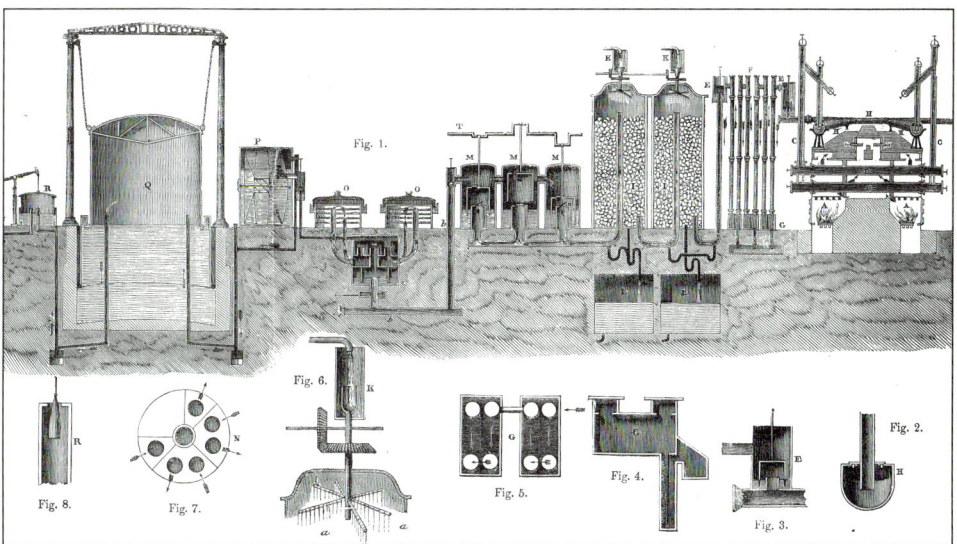

Figure 4.1 **Gas Manufacture**

or compression, or by some combination thereof. There is no reason to believe that in condensing a gas to the liquid state any molecular change to the gas occurs, simply that the molecules of the gas in question are brought very close to each other.

Natural gas, essentially methane in this context, conforms to these general principles and can be liquefied by either cooling or a combination of cooling and compression, but not by compression alone. The reasons why it is impossible to liquefy natural gas by compression alone will be discussed later. Suffice it to say here that the importance of liquefying natural gas is that the transformation from the gaseous to the liquid state reduces the volume by approximately six hundred times, thereby rendering it possible to handle, transport and store a relatively large amount of energy in a more convenient liquid form.

Before discussing processes for liquefying natural gas, it is appropriate to review the historical background, early experimentation and some relevant basic gas laws.

The Work of Robert Boyle

It is difficult to know where to begin, but probably a good starting point as any is the work of Robert Boyle (1627-1691), an Irish physicist, who demonstrated the elasticity of air at Oxford by means of an air pump in 1662.

Boyle discovered that air was not only compressible, but that this compressibility varied inversely with the pressure applied. What Boyle omitted to state was that this relationship only held good if the temperature was kept constant. This relationship between pressure, volume and temperature became known as Boyle's Law in the English speaking scientific world. Some 15 years later, and quite independently of Boyle, Edmé Mariotte (1620-1684), a French physicist, also discovered the same phenomenon as Boyle. Unlike Boyle he stated that this relationship only applied if the temperature remained constant. Because of Mariotte's more precise definition, Boyle's Law is generally known as Mariotte's Law in continental Europe, not without some justification.

The work of Boyle (and Mariotte) was perhaps the first significant step forward in mankind's knowledge of the behaviour of gases.

First Steps in the Liquefaction of Gases

It is uncertain when and to whom the idea that gases could be liquefied first occurred. It is likely, but disputable, that it was Antoine Laurent Lavoisier, a French chemist born in 1743. Lavoisier is famed for his many discoveries and inventions. He was the first to express the idea that matter is neither lost nor created, also for publishing what was probably the first modern textbook on chemistry.

In his opinion, at least part of the earth's atmosphere would liquefy if it was cooled to the temperature of outer space, but he did not attempt to prove this theory.

Lavoisier's work came to an untimely end when he was guillotined in Paris at the age of 51 on 8 May 1794, but not before he had devised a process to make 'water gas', a low calorific value combustible gas composed predominantly of carbon monoxide and hydrogen, improved existing techniques for using combustible gases for street lighting, and had invented a simple type of gasometer.

Lavoisier is now universally recognised as the father of modern chemistry and one of the greatest chemists that has ever lived. His premature death was a tragic loss to the world of science in the eighteenth century.

Most probably Lavoisier's associates, Gaspard Monge (1746-1818) and Jean-Francois Clouet (1751-1801), were the first to succeed in liquefying a gas when they liquefied sulphur dioxide by passing it through a glass tube cooled with a mixture of salt and ice. Around the same time (circa 1790) the Dutchman, Martin van Marum (1750-1837), and his colleague Paets van Troostywk, compressed ammonia and found that at a certain pressure, the volume decreased suddenly and that droplets of liquid ammonia appeared. In 1799, Baron Louis Guyton de Morveau (1737-1816), a French chemist, also liquefied ammonia, by cooling rather than by compression. And in 1805, Northmore compressed various gases up to 15 atmospheres and succeeded in liquefying chlorine.

So by the end of the eighteenth century, it had been proven that at least some gases could be liquefied either by cooling or by compression. These early experiments were underlined and advanced by the work of Sir Humphrey Davy (1778-1829) and his brilliant assistant, Michael Faraday (1791-1867). In 1823, in papers read before the Royal Society in London, Davy described how gases such as chlorine and sulphuretted hydrogen had been liquefied. Encouraged by the work of Thilorier, who condensed carbon dioxide in 1834 on a large scale, Faraday set out to determine the vapour pressure of all gases which could be liquefied at temperatures above minus 110°C, and at pressures up to 50 atmospheres. But his failure, and that of others, to liquefy gases such as methane, hydrogen, oxygen and nitrogen, gave rise to the notion that these gases (and some others) were 'permanent' – a term still in use today, although with qualifications.

The Behaviour of Gases

Based on earlier work by Guillaume Amontons (1663-1705), a French physicist, who demonstrated that a gas changed in volume by the same amount for a given temperature, Jacques Charles (1746-1823), another French physicist, proved and defined this more precisely in 1787. Charles discovered that for each rise or fall in one degree centigrade the volume of a gas expanded or contracted by approximately $1/273$ of its volume at 0°C. This work by Charles became known as Charles's Law.

In 1802, Joseph Gay-Lussac (1778-1850), a French chemist, duplicated Charles's experiments and published his conclusions. Hence Charles's Law is also sometimes called Gay-Lussac's Law. The law, whichever appelation one gives it, states that:

> "The volume of a given quantity of gas is proportional to the absolute temperature if the pressure is held constant".

This is a so-called ideal law, as also is Boyle's Law, as was discovered subsequently by the German-French chemist Henri Regnault (1810-1878).

Regnault was noted for his meticulous measurements and was able to prove that Boyle's Law only applied exactly to an 'ideal' or 'perfect' gas, i.e. where there were no attractive forces between the gas molecules and where the molecules were of zero size.

In fact no known gas has these properties, although hydrogen and helium come very close to being 'perfect' gases. It was not until 1873 that van der Waals worked out a series of complicated gas equations from which the actual behaviour of real gases, as opposed to an ideal gas, could be determined. For his work on gas equations van der Waals was awarded a Nobel prize for physics in 1910.

John Dalton (1766-1844), an English chemist, enunciated his law in 1803. Dalton's Law, which also only holds good for an ideal gas, is that:

> "Each gas of a mixture exerts a partial pressure equal to the pressure which the same mass of the gas would exert if it were present alone in the given space at the same temperature".

Another significant ideal gas law is that based on the hypothesis of the Italian physicist, Amedeo Avogadro (1776-1856), in 1811. Avogadro's Law states that:

> "Under the same conditions of temperature and pressure equal volumes of all gases contain the same number of molecules".

Two other discoverers of ideal laws, relating to the solubility of gases and the vapour pressure of solutions, deserve recognition. They are William Henry (1775-1836), an English chemist, and the French chemist, Francois Raoult (1830-1901). Henry's Law states that:

> "At any specified temperature the amount of gas absorbed by a liquid is proportional to the partial pressure of the gas in contact with the liquid".

While Raoult's Law states that:

> "The partial pressure of solvent vapour in equilibrium with a solution is directly proportional to the mole fraction of the solvent".

Critical Point (of temperature)

Although in 1822 Charles Cagniard de la Tour (1777-1859) proved that when a liquid was heated above a certain temperature it was converted completely into vapour, his experiments received little attention. It was not until 1860 that these and other experiments were repeated by Thomas Andrews (1813-1885), an Irish physical chemist. Andrews took these experiments a stage further and attempted to liquefy the so-called permanent gases by the application of high pressure at various temperatures down to minus 110°C.

Failing, like others, to liquefy them, he turned his attention to the process of liquefying carbon dioxide. He found that carbon dioxide could be liquefied by applying a pressure of 73 atmospheres (1072 psi), provided that the temperature did not exceed 31.1°C. Liquefaction could also be achieved at lower temperatures and with lower pressures, but no liquefaction occurred at temperatures above 31.1°C irrespective of the pressure applied. Andrews determined from his work on carbon dioxide and other gases that there was a critical point, or temperature, above which a gas cannot be liquefied by the application of pressure alone.

Some examples of critical temperatures and pressures for various substances are given in Table 4.1.

The results of Andrews's work were first published in Miller's 'Chemical Physics' in 1863 and formed the subject of his lecture 'On the continuity of the gaseous and liquid states of matter' in 1869. Other accounts suggest that the Russian chemist, Dmitri Mendeléev (1834-1907), made the same discovery a couple of years earlier, but like Cagniard de la Tour his reports went largely unnoticed.

Table 4.1 **Examples of Critical Temperatures and Pressures for Selected Substances**

Substance	Formula	Critical Temperature O°C	Critical Pressure bar
Methane	CH_4	−82.1	45.8
Ethane	C_2H_6	32.2	48.2
Propane	C_3H_8	96.8	42.0
n-Butane	C_4H_{10}	152.0	37.5
n-Pentane	C_5H_{12}	196.6	33.3
Helium	He	−267.9	2.3
Hydrogen	H_2	−239.9	12.8
Nitrogen	N_2	−147.0	33.5
Oxygen	O_2	−118.4	50.1
Carbon Dioxide	CO_2	31.1	72.9
Hydrogen Sulphide	H_2S	100.4	88.9

This highlights one of the difficulties in researching the history of the liquefaction of gases during the eighteenth and nineteenth centuries, namely the conflicting claims as to who discovered what, first and when. The superior communications that existed in European countries, and the abundance of learned societies and published literature, gave European scientists an edge over many of their contemporaries elsewhere. The reality is that several scientists, quite independently of each other, made similar discoveries around the same time, but some received better publicity.

Van Der Waals's Equation

As touched on earlier, in 1873 Johannes van der Waals (1837-1923), a Dutch physicist, published his dissertation in Leiden entitled: 'On the continuity of the liquid and gas states' which refined the earlier work of Andrews. Van der Waals indicated that while the isothermals for temperatures which differ by a very small amount above and below the critical temperature are practically identical in form, the isothermals for the higher temperatures represent the gaseous state only and those for the lower temperatures relate also to the liquid state.

From his studies van der Waals formulated the equation which made him famous and which even today provides the means of calculating very closely, but not absolutely precisely, the physical constants for real gases, as opposed to an ideal gas, on which the various gas laws described above are based.

In layman's terms, van der Waals's equation, which relates the pressure, volume and temperature of real gases, includes two constants which are different for each gas, and recognises, based on actual observations of different gases, the existence of the differing size of molecules and the attractions between them for each gas. Another of his equations determined that by using the temperature, pressure and volume of gas at its critical point, different constants were not needed and this equation would hold good for any gas.

Joule-Thomson Effect

Gay-Lussac in his experiment of 1806, discovered that when a gas under pressure was allowed to expand freely no drop in temperature could be observed.

Nearly 40 years later, James Prescott Joule (1818-1889), an English physicist, repeated and perfected Gay-Lussac's work. More importantly, in conjunction with Sir William Thomson (1824-1907), later to become Lord Kelvin, Joule discovered that under modified experimental conditions, if a gas under high pressure is allowed to expand into a vessel at low pressure, without doing work, but so that no heat is absorbed externally by the expanding gas, then the temperature of the gas will fall. This cooling effect is zero for perfect gases, small for gases at moderate pressure, but substantial for many gases under high pressures.

The results of these experiments have become known as the 'Joule-Thomson Effect'.

Two qualifications deserve mention. First, the cooling effect which accompanies the adiabatic expansion of a gas which does work during its expansion is considerably greater than the Joule-Thomson Effect. Second, the principal practical difficulty which prevents the full realisation of the adiabatic effect is that no material exists which will absolutely prevent any leakage of heat, either inwards or outwards. The material would have to have perfect non-conducting qualities and no capacity for heat.

In order to utilise the Joule-Thomson Effect for a (natural gas) liquefaction process, the gas in question must first be cooled below its so-called inversion temperature which is, however, considerably higher than its critical temperature. Above its inversion temperature, which varies with the pressure applied, the cooling effect disappears and under these conditions expansion will result in a slight heating of the gas.

Figure 4.2 **Gas Standards**

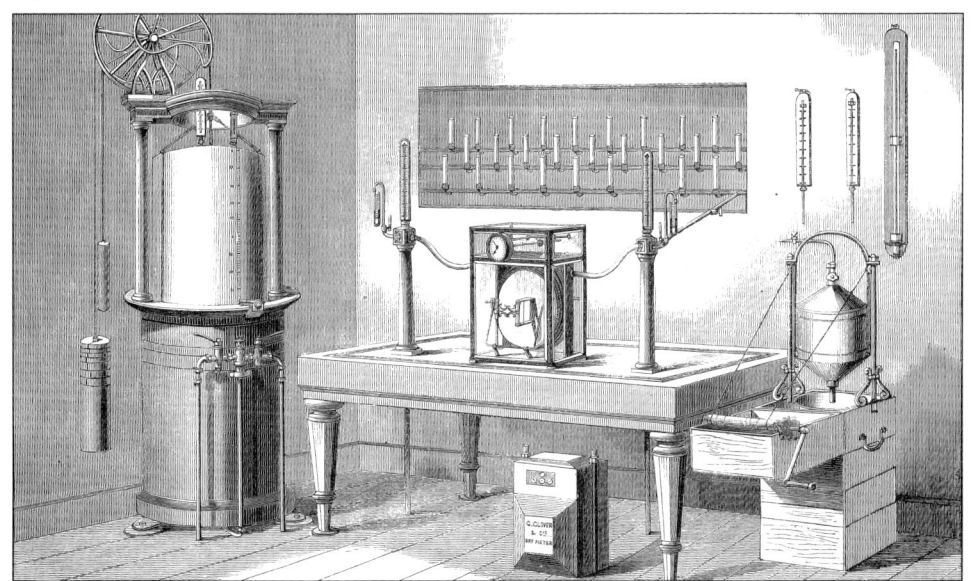

George Glover's patent standard gasometer, meter, and direct transferrer

Liquefaction of Methane by Adiabatic Expansion

Based on the work of Gay-Lussac, Joule and others, Louis Cailletet (1832-1913), a French ironmaster and physicist, first liquefied acetylene by adiabatic expansion, essentially the Joule-Thomson Effect.

He then proceeded to attempt to liquefy the so-called permanent gases by similar means. On 2 December 1877, Cailletet liquefied oxygen at a pressure of 300 atmospheres and at a temperature of minus 29°C in a bath of sulphur dioxide boiling under reduced pressure, with the oxygen then being allowed to expand by releasing the pressure. He followed this by successfully liquefying methane and carbon dioxide with the same procedure and apparatus.

Page 45

Liquefaction of Methane by Mechanical Refrigeration

At the same time as Cailletet was conducting his experiments, Raoul Pictet (1846-1929), a Swiss physicist, also liquefied oxygen, nitrogen, carbon dioxide and methane.

Pictet achieved his first success by the continuous liquefaction and evaporation of sulphur dioxide. By condensing sulphur dioxide under a pressure of three atmospheres and evaporating the resultant liquid in vacuo he reached a temperature of minus 65°C. At this temperature he condensed carbon dioxide, and by evaporation obtained a temperature of minus 130°C. Having obtained this temperature, he cooled oxygen, which on release under a pressure of 200 atmospheres liquefied.

The results of both Cailletet and Pictet's work were published almost simultaneously. Which of these two was actually the first to liquefy oxygen remains in dispute. However, the balance of evidence suggests that for methane, Cailletet was probably the first to condense this gas in December 1877. If Cailletet is to receive the credit for being the first to produce liquid methane, it was Pictet who achieved the more practicable approach to gas liquefaction in a potentially commercial sense.

Liquefaction of Methane by the Cascade Process

First processes for the continuous production of a liquefied gas, initially oxygen, were devised by Sir James Dewar, (1842-1923), a Scottish scientist, and Heike Kamerlingh Onnes (1853-1926), a Dutch physicist, based on the original work of Pictet. Both Dewar and Onnes utilised a series of gases of progressively lower boiling points to achieve this. This gave rise to the so-called cascade process of liquefaction.

However, perhaps Dewar is best known for inventing in 1872 the double-walled vacuum flask for the storage of liquefied gases. Onnes in turn is credited with being the first to liquefy helium, the last of the permanent gases, which he achieved in July 1908 using Dewar's techniques, and in so doing reached a temperature within one degree above absolute zero.

In conclusion, and at the risk of repetition, omitted from this account is the work of many other scientists who have contributed to the study of the behaviour and properties of gases including Witkowski, Ramsay, Young, Clausius and Berthelot. While others like Cope, Lewis, Weber, Luke and Kay devised methods to calculate deviations from the perfect gas laws for known mixtures of hydrocarbon gases as comprise various natural gases.

Some readers may wish to research the literature to gain a more comprehensive and better balanced account of this subject than is given here.

Chapter 5

GAS TREATMENT AND PROCESSING

To state the obvious, the extent to which natural gas has to be treated and processed, and the processes used, will depend upon its original composition and the proposed end-use. A further consideration may be any added-value opportunities to extract for separate sale certain constituents that may be present.

For the purposes of this chapter it is assumed that the main objective is to produce a marketable quality of gas for sale to a gas utility, that the gas contains unacceptable amounts of water, carbon dioxide and sulphur compounds, also sufficient quantities of gas liquids to justify/necessitate their extraction. Liquefaction and the conversion of gas to other products are discussed in Chapters 6 and 15 respectively.

Gas Liquids and Sulphur Extraction: Business Scope

Excluding the Soviet Union and China for which reliable data are not available, there were nearly 1500 gas processing plants in service throughout the rest of the world in 1990 of which approximately half are in the United States; Texas alone has 300. The total gas capacity of these plants amounts to over 156 billion cubic metres (Bcm), although actual gas throughput in 1990 was only 94 Bcm. Out of a total production of some 3.5 million barrels per day (b/d) of gas liquids from natural gas in 1990, some 700000 b/d was ethane, 400000 b/d propane, 240000 b/d butanes, 150000 b/d LPG mixtures and the balance pentanes and heavier fractions. These figures exclude the substantial quantities of gas liquids extracted in field separators in the course of producing crude oil.

Sulphur recovery is also a significant business in some countries, particularly in Canada which in 1990 produced nearly 5.8 million tonnes, or one-third of a total world production of 17 million tonnes of petroleum-derived sulphur.

Dehydration

Water vapour is probably the commonest undesirable impurity in natural gas. It is not the water vapour itself that is objectionable but rather the liquid or solid phase which may precipitate when the gas is compressed or cooled. Moreover, under certain conditions of temperature and pressure, water vapour and hydrocarbons can form hydrates. Liquid water accelerates corrosion and accumulates at low points in pipelines and equipment reducing their capacity, while ice and hydrates can plug valves and lines.

There are four basic methods by which water vapour may be reduced to acceptable limits:

i) *compression to a higher pressure with subsequent cooling;*

ii) *cooling below initial dew point, commonly known as low-temperature separation;*

iii) *adsorption with dry desiccants; and,*

iv) *absorption with liquids.*

The selection of any one or some combination of the above methods, will depend upon the extent to which water vapour and any other components need to be extracted, the natural pressure of the gas when it enters the plant and the send-out pressure required, cost considerations, etc.

Use of the first method, COMPRESSION AND COOLING, is rather limited as the degree of dehydration achieved is not always sufficient at a given compression. Further drying can be achieved by applying more cooling, but usually it is cheaper

Figure 5.1 **Gas Drying Unit, Den Helder, The Netherlands**

Courtesy of Nederlandse Aardolie Maatschappij B.V.

to dry the gas by other means before compression thereby avoiding expensive cooling facilities necessary to condense all vapours that have to be removed.

The second method, LOW-TEMPERATURE SEPARATION, is more common, particularly where a substantial pressure differential exists between the gas inlet pressure and the required send-out pressure as the necessary low temperatures can be achieved by expanding the gas. A complication here is that at high pressures, and in the presence of free water, hydrates may be formed with ethane, propane and, if present, hydrogen sulphide and carbon dioxide, at temperatures above the melting point of ice and at pressures lower than for methane. This can be overcome by injecting glycol into the gas stream which prevents hydrates forming and allows the process to be operated at lower temperatures.

An example of this process (based on the Joule-Thomson effect) is the treatment of Dutch Groningen gas. Gas from wells is passed through a high pressure liquid separator and, before being expanded in the low-temperature separator, is cooled in two stages to around $20°C$ which is close to the hydrate formation point. Glycol is injected into the gas stream which is then expanded reaching a temperature of about $-12°C$. Any solids which may be formed in the bottom of the low-temperature separator are melted by a coil through which hot glycol from the glycol regenerator is passed. Hydrocarbons, water and glycol precipitated from the gas in the low-temperature separator flow to a separator where the hydrocarbons are taken off. The glycol-water is sent to a concentrator where the glycol is regenerated and recycled.

Low-temperature separation has the merit of hydrocarbon recovery in addition to dehydration. Its limitation is that it requires a gas inlet/send-out pressure differential of about 80 bar or more to achieve the necessary dew point depression. Dew point is the temperature at which a vapour, contained in a closed vessel under the given pressure, will form a first drop of liquid on the subtraction of heat.

ADSORPTION processes mostly use silica or alumina based materials which have been prepared to have a very large surface area per unit weight, since the amount of substance (i.e. water) adsorbed is directly related to the surface area available for adsorption.

The process is relatively simple. For a continuous gas flow, two columns containing the dry desiccants are needed – one adsorbing while the other is being regenerated. Inlet gas is first passed through a separator before it is fed to the top of the operating column where it passes through the desiccant bed to the bottom. Hydrocarbon and water vapours are adsorbed by the bed. When the bed is completely saturated with water, the gas stream is switched to the second column while the bed in the first column is heated with hot regenerator gas to vaporise the entrained water. Regenerator gas and water vapour are passed to a regeneration gas cooler, where the water condenses and is separated from the gas stream. Once the bed of the first column has been dried, it is cooled down before being switched back into service. The cycle is then repeated.

Because hydrocarbons are also adsorbed in this process, it can be used for the extraction of heavier hydrocarbon fractions. Dew points down to about -35°C can be achieved and gas temperatures up to about 40°C are acceptable. Disadvantages include a high pressure drop over the process, the desiccants are sensitive to poisoning by, for example, heavy oils that may become entrained in the gas, and initial investment is relatively high.

ABSORPTION processes with liquids mainly use diethylene glycol (DEG) and triethylene glycol (TEG), although other liquids, e.g. sulphuric acid, glycerol, etc., are sometimes used. Both DEG and TEG have good stability at high temperatures, high hygroscopic capacity and low vapour pressures. Dew point depressions of up to 15°C with DEG and up to 30°C with TEG can be achieved. For dehydrating gases containing carbon dioxide and hydrogen sulphide, glycol-amine mixtures are used.

Following primary separation of free liquids in an inlet separator, the gas stream enters the bottom of the contactor tower and bubbles upwards through a series of trays containing glycol which flows from top to bottom absorbing water vapour on its way down. Glycol exiting from the bottom of the contactor is filtered and then fed to a regenerator where the water is stripped and the glycol pumped back to the contactor.

Glycol processes are less efficient and less flexible than adsorption processes but much cheaper.

In general, the most widely used processes today are either low-temperature separation, by which most water and certain hydrocarbons are removed as liquids, or separation to remove any free liquids followed by glycol contacting to remove water vapour.

Hydrocarbon Extraction

The extent to which ethane and such other heavier hydrocarbons as may be present are removed from pipeline gas (LNG is discussed in Chapter 6) will be conditioned primarily by the buyer's gas quality specification and the need to avoid propane and heavier fractions condensing out in pipelines by cooling and/or compression. An additional consideration is, of course, the added value opportunities that may exist for extracting such components. The removal of acid gases and/or inerts, if present, is discussed later.

Some natural gases have relatively few or negligible amounts of propane and heavier fractions which are not economic nor necessary to extract, while other gases, particularly associated gas, may be rich in such components. In the latter case, removal of such components to the necessary level may be carried out in one plant or in two stages in separate plants.

Figure 5.2 **St. Fergus Gas Plant, Scotland**

Two modules are on the left and the fractionation column is in the right background.

For example, most gases from fields in the UK sector of the northern North Sea are very rich in gas liquids. They are first piped to an onshore beach plant at St. Fergus in Scotland – see Figure 5.2 – where the liquids are removed so as to produce a gas of pipeline quality.

A cocktail of extracted ethane and heavier fractions is then further processed at another plant at Mossmorran over 200 km away to produce ethane as a feedstock for an ethylene cracker and a range of marketable liquid products. This contrasts with the situation for southern North Sea non-associated gas which is relatively lean and requires only limited treatment in one plant. For illustrative purposes the first example has been taken.

At St. Fergus a variety of gases of different compositions from different fields is received. They are first passed through a slug catcher, essentially a matrix of pipes – see Figure 5.3 – which removes liquid 'slugs' from the feed gas and smooths out the liquid and vapour flows. The rich 'wet' gas is then passed to an inlet separator which separates the remaining liquids from the gas stream.

The gas stream is then sent to a gas dehydration unit and the liquids to a liquid dehydrator where water is removed. Both streams are then cooled to $-100^{\circ}C$ by heat

Figure 5.3 **Slug Catcher, St. Fergus Gas Plant**

exchange and by two stages of turbo expansion before being fed to a de-methanizer column where methane is stripped. Methane from the top of the de-methanizer is then compressed, measured and sent off to the pipeline system.

Part of the product from the bottom of the de-methanizer is fed to a product fractionation unit where a portion of the ethane is recovered. This is then blended with the methane from the de-methanizer to ensure that the outgoing pipeline gas meets the buyer's quality specification. The remaining ethane and heavier fractions

Figure 5.4 **Mossmorran Fractionation Plant**

One of the distillation columns

from the de-methanizer and product fractionation units are then pumped at 70 bar through a 16 inch, 220 km pipeline to Mossmorran for further processing.

At Mossmorran – see Figure 5.4 – the feed from St. Fergus is passed sequentially to three distillation columns for fractionation. The first is a de-ethanizer which removes ethane as the feed for an adjacent ethylene cracker (see Chapter 15, Figure 15.4). The second column, a de-propanizer, separates out propane which is treated to remove H_2S before being cooled, liquefied and piped to refrigerated storage. The third column, a de-butanizer, removes and liquefies butane which is piped to separate refrigerated storage. The remaining gas liquids, pentanes and heavier

Page 53

components, from the bottom of the de-butanizer are fed to conventional type oil product storage for sale. Refrigerated propane and butanes are exported by ship via a nearby loading terminal.

In reality, the processes involved are rather more complex than this simplified account suggests and result in the production of high quality pipeline gas, ethane, propane, butanes and condensates.

Acid Gases

Apart from chemically inert nitrogen, most gaseous impurities present in sizeable amounts in natural gas, i.e. carbon dioxide and sulphur compounds, have an acidic reaction. Hence their generic name of 'acid gas'.

In earlier years H_2S was often combusted to SO_2 and vented to the atmosphere, although processes for the removal of acid gases have been available for more than 60 years. Venting is no longer permitted in many countries other than for quantities sufficiently small enough to comply with environmental standards.

Other impurities include CO_2, carbonyl sulphide (COS) and mercaptans. CO_2 is not usually removed to the same level as H_2S as it is less objectionable, more expensive to remove, and in some cases is a desirable component, e.g. for gas conversion processes – see Chapter 15. Most natural gases do not contain significant volumes of COS but if they do, COS can be absorbed in a great many solvents. Although mercaptans (principally methyl mercaptan, CH_3SH) are present in most sour natural gases, the majority of processes used remove mercaptans to acceptable limits.

There are many commercially proven proprietary processes available for the treatment of acid gases. They can be divided broadly into physical solvent and chemical solvent processes, and there are sub-divisions of these categories, for example chemical solvents can be divided into alkanolamines and alkaline processes.

Selection of the preferred process will depend upon many factors including the pressure of the feed gas, the amount and combination of acid gases in the gas, the specification required of the treated gas, whether separate recovery of sulphur is required/justified, allowable hydrocarbon losses, etc.

Each main type of process has its advantages with chemical solvent processes generally being preferred, although they use more energy. Physical solvent processes tend to be used where gas pressures are high and for gases with low concentrations of propane and heavier hydrocarbons. Some processes combine chemical and physical solvents and are particularly attractive when the partial pressure of the acid gases is moderate.

As it is impractical to cover the whole range of processes available, four representative processes have been selected for illustrative purposes, three of which have been developed by and are Shell proprietary processes. Shell also has its own design of the Claus process – all four are widely used.

Typical Flow Sequence

A typical flow sequence for a sour gas involving sulphur recovery would be initial separation of water and liquid hydrocarbons from the feed gas in an inlet separator; absorption of trace amounts of propane and heavier hydrocarbons from the gas in an absorption unit; removal of acid gases in a Sulfinol unit with the 'sweet' gas being sent to a dehydration unit for the further removal of water to meet pipeline gas quality specifications and, depending on its composition, the dried sweet gas may then be processed to extract ethane, propane and butanes for separate sale.

Acid gas from the Sulfinol unit is then sent to a Claus sulphur extraction plant and converted to elemental liquid sulphur. Tail gas from the Claus plant may be further processed in a SCOT plant where it would be converted to H_2S and recycled to the Claus plant for conversion to liquid sulphur - the Sulfinol, Claus and SCOT processes are described later. Liquid sulphur is then formed into solid pellets. Any remaining acid gases not recovered in the SCOT plant are incinerated and vented.

Ancillary processes would involve treating extracted water to an acceptable quality, or injecting it into deep underground formations, and removing any remaining gases from extracted condensate in a condensate stabiliser to produce a marketable product.

The Shell Adip Process

This process removes H_2S down to about 5 ppm. It also partly removes CO_2 and COS if present. It is based on regenerative absorption by a solvent amine, usually an aqueous di-isopropanolamine (DIPA) solution. As this solution requires a fairly long residence time it is technically more suitable for liquefied gases than gas, for which the Sulfinol process is more often used.

Figure 5.5 **Shell Adip Process Flow Scheme**

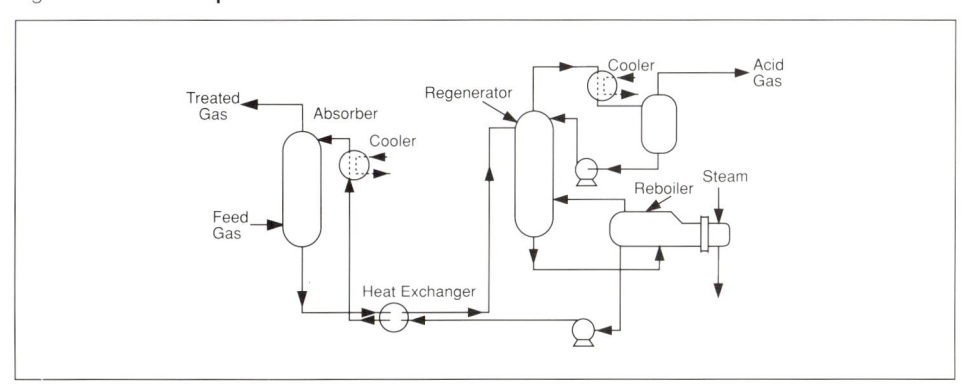

The feed gas stream is contacted counter-currently in an absorption or extraction column with DIPA. The regenerated amine solution is introduced at the top of the absorption column and leaves at the bottom. The rich solution is heated by heat exchange with the regenerated solution from the bottom of the regenerator and is fed to the regenerator, where is is further heated and freed of the acid gases with steam.

After flowing through the heat exchanger, the regenerated solution is cooled by heat exchange with air or water and then introduced into the absorption column. Subsequently, the acid gases removed from the regenerator are cooled with air or water, so that the major part of the water vapour they contain is condensed. The sour condensate is reintroduced into the system as a reflux. In this way, all H_2S is contained in the acid gas stream. Figure 5.5 shows the flow scheme where a high degree of H_2S removal is required.

As the corrosivity of DIPA is low, the plant can be built with carbon steel. Steam consumption is lower than for most similar processes using different solvents. Absorber temperatures and pressures typically range around $40°C$ and 25 bar.

The Shell Sulfinol Process

The Sulfinol process combines chemical and physical solvents to form a mixed solvent in a process which is more versatile than either of the two individually. The mixed solvents behave like a chemical solvent at low pressures and like a physical solvent at high pressures. The Sulfinol process is one of the most widely used processes for the removal of H_2S, CO_2, COS, mercaptans, etc., from sour gas streams.

The process is very similar to the ADIP process, the main difference being that it uses an organic solvent, Sulfolane (tetrahydrothiophene dioxide), mixed with a secondary or tertiary alkanolamine and water. Simultaneous physical and chemical absorption at high and low partial pressures respectively is provided by the Sulfinol solvent. Regeneration is accomplished by release of the acidic constituents at near atmospheric pressure and at somewhat elevated temperatures. As shown in Figure 5.6, a flash of the fat solvent may be used, also flashed gases may be recycled to the absorber.

The advantages of the Sulfinol process are that the Sulfolane component of the solvent is chemically and thermally very stable and degradation is negligible. Solvent circulation is reduced due to its high acid gas solubility with lower energy required for regeneration. Treated gas meets pipeline gas quality specifications for acid gases. Corrosion rates are also very low – there are other advantages.

The Sulfinol process operates typically at absorption pressures of some 60 bar and temperatures of about $40°C$. H_2S and COS can be removed to less than 7 ppm (volume) and CO_2 to less than 2 per cent. The principal disadvantage is that the

Figure 5.6 **Shell Sulfinol Process Flow Scheme**

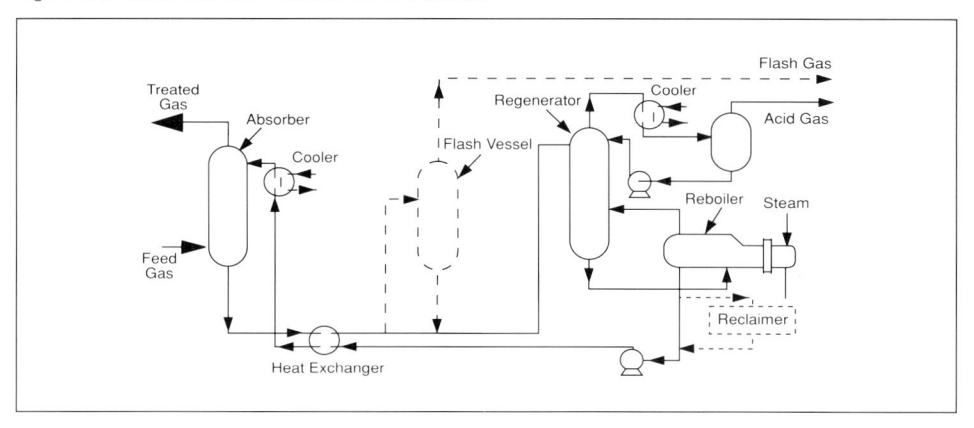

solvent absorbs aromatic hydrocarbons that may be present in the feed gas stream. However, these can be removed easily by treating the acid gas stream by charcoal adsorption before the acid gas is fed to a sulphur recovery plant.

The Claus Process

Claus type processes are based on partial combustion of H_2S to SO_2 at about 1200 to 1400°C and the further catalytic reaction of H_2S and SO_2 to form elemental sulphur, viz:

$$2H_2S + SO_2 \longrightarrow 3S + 2H_2O$$

In the thermal stage part of the H_2S is burnt with air to form sulphur dioxide such that the molecular ratio of H_2S to SO_2 is two. Under the prevailing temperature and pressure conditions, H_2S and SO_2 react to form sulphur (S_2) with yields of up to about 70 per cent. After cooling the process gas, condensation and removal of the sulphur in a waste heat boiler, more sulphur is formed in one or several catalytic stages. Each catalytic stage consists of a process gas reheat facility and a reactor filled with activated alumina followed by a sulphur condenser.

The process is exothermic, and if little CO_2 is present the combined gas treating and sulphur recovery units are at least self-sufficient in energy terms. Equilibrium limits sulphur recovery to about 95 to 97 per cent. The cost of the process and sulphur emissions increase with CO_2 concentrations in the feed. Excessive hydrocarbons in the feed can result in black and unsaleable sulphur.

The tail gas from a Claus sulphur recovery plant is a mixture of H_2S, SO_2, etc., that remains after condensing the sulphur. It may be combusted to SO_2 and vented if the quantities are small enough to meet environmental standards, or further processed in a SCOT unit where H_2S is selectively removed and recycled to the Claus plant. Selectivity is essential to avoid CO_2 build up.

Liquid sulphur from the Claus plant is usually formed into solid pellets in a prilling tower for ease of storage and subsequent handling.

The Shell Claus Off-gas Treating Process (SCOT)

The SCOT process is usually combined with a sulphur recovery plant, e.g. a Claus unit, to remove virtually all remaining sulphur components from the off-gas. It consists essentially of a reduction unit and an alkanolamine absorption unit.

In the reduction unit all sulphur compounds and free sulphur present in the Claus off-gas are completely converted into H_2S over a cobalt/molybdenum catalyst at $300°C$ in the presence of H_2 or a mixture of H_2 and CO. The reducing gas can either be supplied from an outside source or generated by substoichiometric combustion in the direct heater which heats the process gas to the reactor inlet temperature. The reactor effluent is cooled subsequently in a heat exchanger and a cooling tower. Water vapour in the process gas is condensed and the condensate is sent to a sour water stripper.

Figure 5.7 **Shell Scot Process Flow Scheme**

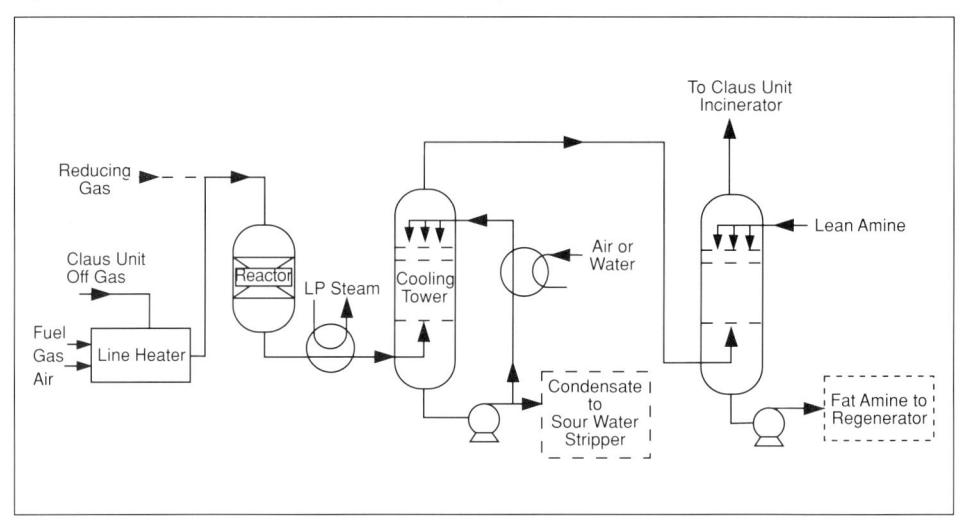

The cooled gas, normally containing up to 3 per cent by volume of H_2S and up to 20 per cent by volume of CO_2, is counter-currently washed with an alkanolamine solution in an absorption column specially designed to absorb almost all H_2S but relatively little CO_2. The treated gas from the absorption column contains only traces of H_2S and is burned in a standard Claus incinerator. The concentrated H_2S is recovered from the rich absorbent solution in a conventional stripper and is recycled to the Claus unit.

In practice, an overall yield of more than 99.8 per cent sulphur recovery can be obtained with the SCOT process.

Figure 5.8 **A Scot Plant**

The above SCOT plant is part of Shell Canada Resource's Waterton gas plant in Alberta. The Waterton plant can produce over 2500 tonnes of sulphur per day from natural gas.

Carbon Dioxide Removal

Carbon dioxide is inert at ambient temperatures and acidic in the presence of water. As the former lowers the calorific value and the latter can cause corrosion in pipelines and related equipment, most pipeline gas quality specifications require the removal of carbon dioxide to 2 per cent or less by volume.

Hydrogen sulphide is frequently present in gases containing carbon dioxide. Many processes for the removal of H_2S also almost completely remove CO_2 as they are not selective. The Sulfinol process is a case in point. Equally, some processes are not suitable if H_2S is present, or do not remove CO_2 down to the usual levels required, e.g. the ADIP process. CO_2 removal processes can be divided into non-regenerative and regenerative processes and the latter into physical adsorption and chemical absorption processes.

While all processes for the removal of acid gases are discrete to varying degrees, the illustrative descriptions given above of regenerative processes are considered adequate as regards CO_2 removal for the purposes of this book. Non-regenerative processes are normally only economic if only small amounts of carbon dioxide, but no H_2S, have to be removed. They have the obvious disadvantages of being non-regenerable with the consequential need for safe disposal of the spent solvent. Basically, they work on the principle of the gas stream being treated counter-currently in a column filled with circulating aqueous caustic soda. Fresh solvent is continuously injected as the spent solvent is bled off for disposal.

Denitrogenation

As nitrogen is an inert and stable gas it is not normally extracted if present in small quantities, especially if the gas is relatively free of other inerts and acid gases. Exceptionally, it may even be left in the gas when present in relatively large quantities, as is the case with Dutch Groningen gas which contains over 14 per cent nitrogen but less than one per cent carbon dioxide and only a trace of hydrogen sulphide.

The two main reasons for nitrogen removal are to ensure that the resultant gas is easily interchangeable with other gases containing less or no inerts and/or because it has no heating value it increases the cost per unit of useful energy transported to end-consumers. This can be an important economic consideration if the gas has to be transported over long distances, with the additional cost of transportation being weighed against the cost of extraction.

Where nitrogen extraction is deemed to be necessary or desirable, it is usually removed by low temperature distillation. The feed gas is first dried and then passed through a heat exchanger where it is cooled against waste nitrogen and the treated gas streams. The partly liquefied gas stream is then reboiled and partly refluxed by a closed methane loop in a distillation column. Methane is condensed in the reboiler under a pressure slightly above the column pressure. The liquefied methane is then passed through liquid subcoolers and evaporated in the top part of the column at lower pressures. Heat is absorbed in that part of the column whereby reflux is generated. If more reflux is required, this can be generated either by expanding the top product through an expander thereby producing more cold or with a closed nitrogen cooling cycle. Reflux is a part, if the top product is in the liquid state, or all, if in the vapour phase, of the condensed top vapour of a column, which is returned to the top of the column. The purpose is to create an extra downward flow of liquid. If properly applied this liquid acts as an absorbing agent for heavier components which are thus rejected from the top product.

The top product usually contains sufficient methane (15 to 20 per cent by volume) for generating the energy required in a gas turbine for the separation process.

The bottom product, essentially liquid methane, is boosted to pipeline gas send-out pressure and heated in the main heat exchanger against the feed gas, before being sent out.

Mercury

Few natural gases contain mercury and then only in very small amounts. Its removal is only essential if the gas is likely to be in contact with aluminium and aluminium alloys. For further details see Chapter 6.

Helium

Helium occurs naturally in minute proportions in the earth's air. At ordinary temperatures and pressures, helium conforms closely to the laws of an ideal gas. The only known material from which helium can be extracted economically is helium-bearing natural gas. Those natural gases which contain helium, typically less than one per cent by volume, usually originate from reservoirs located over buried granite. As helium in such small concentrations causes no problems in gas processing, transportation and use, more often than not it is left in the gas unless there are worthwhile opportunities to extract it for separate sale.

The extraction of helium requires additional processing facilities specifically for that purpose. Basically, they involve a combination of pressure and cooling with nitrogen. In one example process, there are others, the feed gas first enters a knock-out drum to remove liquids and is then dried. The gas is then compressed to about 45 bar, cooled in stages to around $-100^{\circ}C$, reduced in pressure and delivered to a nitrogen-methane fractionator. Product from the top of the fractionator is a mix of nitrogen and helium, and from the bottom methane enriched gas.

The nitrogen-helium stream is fed to a nitrogen condenser heat exchanger where it is cooled to about $-190^{\circ}C$. Helium is then separated from the nitrogen by flashing in drums at an initial pressure of about 25 to 30 bar, followed by a secondary recovery flash at 5 bar, giving a helium of about 95 per cent purity. Higher purity of up to very nearly 100 per cent can be achieved by passing the helium through purifiers containing activated charcoal cooled by liquid nitrogen.

Helium is then compressed and stored in special containers at pressures in excess of 150 bar for transport by road or rail.

Part of the nitrogen is used in the process, but the greater part is warmed up to ambient temperatures and vented to atmosphere if there is no other use for it.

Chapter 6

BASE-LOAD LIQUEFIED NATURAL GAS PLANTS

Natural gas was first liquefied on a practical scale by the United States Bureau of Mines in 1917. In this instance the purpose was to separate helium from natural gas to obtain helium for airships and not to liquefy natural gas per se.

The world's first commercial liquefaction plant was an LNG peak shaving facility built in Cleveland, Ohio, in 1941. It operated successfully until 1944 when a tank containing LNG failed – see Chapter 7 for further details. Although this accident was a set back for the LNG industry, it at least ensured that major improvements in materials, design, construction and operating techniques were perfected and introduced.

There are now well over 60 LNG peak shaving plants in operation most of which are in North America, but there are a number in Europe and elsewhere: the lessons of Cleveland were learnt and implemented.

As LNG peak shaving plants are almost invariably based on feed gas of marketable quality from transmission systems and distribution grids which is liquefied during periods of low demand, they are, by definition, smaller with fewer processing stages than a base-load LNG plant. Accordingly, in order to cover the full range of activities involved in liquefying natural gas, the remainder of this chapter is devoted to the main components and processes employed in base-load plants, some of which will also be applicable to peak shaving plants.

Operational Base-Load Plants

There are twelve (or fourteen if one regards the Skikda complex as three separate plants) operational, base-load LNG export plants in the world today – see Table 6.1. The first to come into commercial service, in 1964, was the then called CAMEL plant (subsequently designated GL4-Z) at Arzew, Algeria, built specifically for the supply of LNG to England and France. As the story of this project is well documented in the literature, it will not be repeated here.

As one might expect, the design of each plant is unique reflecting technological developments over time as well as different feed gas compositions, local site conditions and the design and process preferences of the plant owners. In addition, design and process selection have been and will continue to be influenced by differences and trade-offs between capital, operating and maintenance costs, fuel efficiency considerations and the like.

Table 6.1 **Operational Base-Load LNG Plants**

Location	Start Up	No. of Trains	Process	Engineering Contractors	Construction Contractors	Capacity mtpa
Arzew GL4-Z Algeria	1964	3	Technip Cascade	Technip & Pritchard	Technip & Pritchard	1.0
Kenai Alaska	1969	1	Phillips Cascade	Bechtel	Bechtel	1.5
Marsa El Brega Libya	1970	2	APCI-MCR	Bechtel	SNAM	1.5
Skikda GL1-K Algeria	1972	3	TEALARC	Technip	Technip	3.0
Lumut Brunei	1972	5	APCI-MCR	JGC & Procon	JGC & Procon	5.5
Das Island Abu Dhabi	1977	2	APCI-MCR	Bechtel & Chiyoda	Bechtel	2.5
Badak Indonesia	1977 1983 1990	2 2 1	APCI-MCR APCI-MCR APCI-MCR	Bechtel Bechtel Chiyoda	Bechtel Bechtel Chiyoda	4.0 4.0 2.0
Arun Indonesia	1978 1984 1986	3 2 1	APCI-MCR APCI-MCR APCI-MCR	Bechtel Chiyoda JGC	Bechtel Chiyoda JGC	4.5 3.0 1.5
Bethioua GL1-Z Algeria	1978	6	APCI-MCR	Bechtel & Chemico	Bechtel & Chemico	9.0
Bethioua GL2-Z Algeria	1981	6	APCI-MCR	Kellogg	Kellogg	9.0
Skikda GL2-K Algeria	1981	1	PRICO	Pritchard & Kellogg	Pritchard & Kellogg	1.5
Skikda GL3-K	1982	2	PRICO	Kellogg	Kellogg	3.0
Bintulu Sarawak	1983	3	APCI-MCR	JGC & Kellogg	JGC & Kellogg	8.0
Withnell Bay Australia	1989	3	APCI-MCR	JGC, Kellogg & Raymond	JGC, Kellogg & Raymond	7.0

Notes to Table 6.1

i) *Start up years are when the first cargo of LNG from the first completed train was loaded and shipped.*

ii) *APCI-MCR is Air Products & Chemicals Inc. Multi-Component Refrigerant process and is a combination of a pure refrigerant cascade process and a mixed refrigerant cycle. Technip Cascade and Phillips Cascade are pure refrigerant cascade processes. TEALARC and PRICO are mixed refrigerant cycle processes.*

iii) *Engineering and Construction Contractors are the principal contractors. In most instances they would have received project specification, technical advice and support from the plant owners, or their appointed technical adviser, also from a variety of specialist sub-contractors. Not shown are those cases where a separate main contractor was appointed to design and/or construct the cryogenic storage.*

iv) *Capacities indicated (rounded off to the nearest 0.5 million tonnes of LNG p.a.) are approximate and generally relate either to long term contractual obligations or to the original designed output. Several plants after debottlenecking/modification are now able to produce larger quantities of LNG on a regular basis e.g. Bintulu. Conversely, some North African plants have yet to operate continuously at their design capacity rate due, inter alia, to the collapse of contractual sales agreements with certain customers. The third train of the Australian plant is still under construction.*

v) *Not shown in Table 6.1 are a number of new trains currently under construction or about to be constructed at existing plants, notably for the Abu Dhabi, Indonesia (Badak) and Malaysia (Sarawak) projects.*

Plant Components

Base-load LNG plants generally comprise the following main components or units – see also Figure 6.1.

- *Gas receiving and metering*
- *Acid gas removal and, if appropriate, sulphur recovery*
- *Dehydration*
- *Mercury removal, if necessary*
- *Heavy hydrocarbon separation*
- *Fractionation*
- *Liquefaction*
- *NGL treatment*
- *LNG and NGL storage*
- *LNG and, if applicable, NGL ship loading facilities*
- *Main utilities*

Collectively these units are known as a 'train'. With the exception of the Kenai and Skikda GL2-K plants, all operational plants consist of more than one train. These trains are operated in parallel and shut down sequentially for periodic maintenance and repair. Provided the plant is well designed and constructed, properly maintained and refurbished as necessary, experience to date suggests that it may have a useful operating life of 40 or 50 years, possibly longer.

In addition to these components, and depending on the location of the plant, a supporting infrastructure of living accommodation, roads, hospital, school, recreational facilities, possibly a harbour and an airstrip, may have to be provided if such facilities are not available or are inadequate. For a greenfield site, such facilities can cost several hundreds of millions of dollars, a substantial proportion of which may have to be borne by the LNG project company.

Figure 6.1 **Schematic LNG Plant: Main Components**

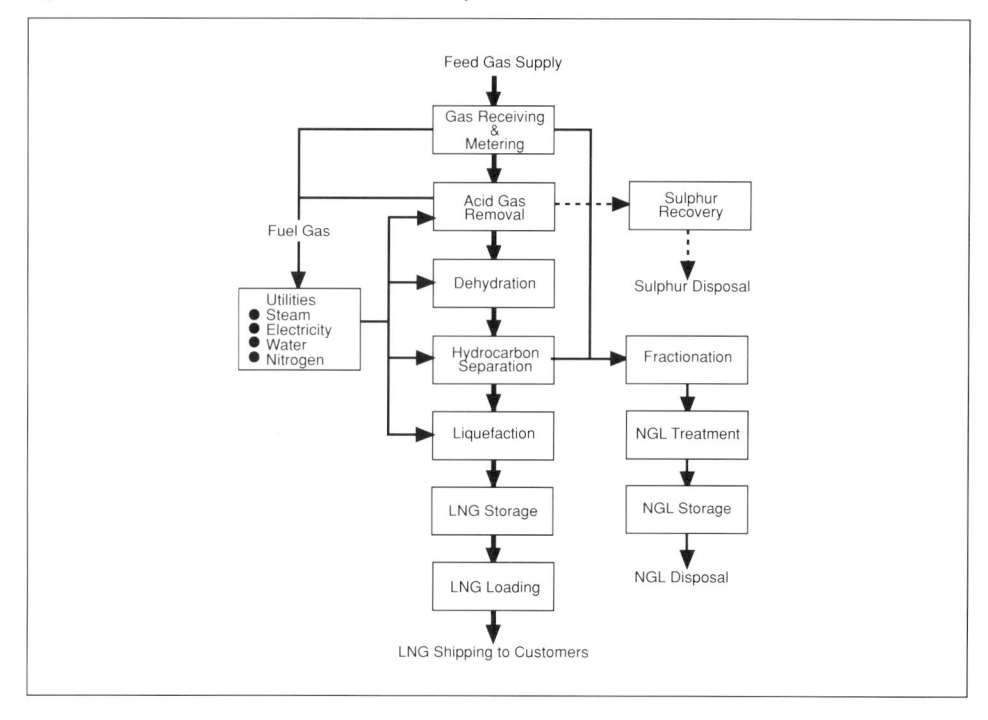

```
                        Feed Gas Supply
                              │
                              ▼
                     ┌─────────────────┐
          ┌─────────▶│ Gas Receiving   │
          │          │       &         │
          │          │    Metering     │
          │          └─────────────────┘
          │                   │
          │                   ▼
          │          ┌─────────────────┐       ┌──────────────┐
          │     ┌───▶│   Acid Gas      │─ ─ ─ ▶│   Sulphur    │
Fuel Gas  │     │    │   Removal       │       │   Recovery   │
          │     │    └─────────────────┘       └──────────────┘
          │     │             │                        ┊
          │     │             ▼                        ▼
          │     │    ┌─────────────────┐       ┌──────────────┐
          │     │    │   Dehydration   │       │Sulphur Disposal│
 ┌────────┴──┐  │    └─────────────────┘       └──────────────┘
 │ Utilities │  │             │
 │ ● Steam   │  │             ▼
 │ ● Electricity │   ┌─────────────────┐       ┌──────────────┐
 │ ● Water   │──┼───▶│  Hydrocarbon    │──────▶│ Fractionation│
 │ ● Nitrogen│  │    │   Separation    │       │              │
 └───────────┘  │    └─────────────────┘       └──────────────┘
                │             │                        │
                │             ▼                        ▼
                │    ┌─────────────────┐       ┌──────────────┐
                └───▶│  Liquefaction   │       │ NGL Treatment│
                     └─────────────────┘       └──────────────┘
                              │                        │
                              ▼                        ▼
                     ┌─────────────────┐       ┌──────────────┐
                     │   LNG Storage   │       │  NGL Storage │
                     └─────────────────┘       └──────────────┘
                              │                        │
                              ▼                        ▼
                     ┌─────────────────┐       ┌──────────────┐
                     │   LNG Loading   │       │ NGL Disposal │
                     └─────────────────┘       └──────────────┘
                              │
                              ▼
                    LNG Shipping to Customers
```

Gas Receiving and Metering

In this unit the feed gas for the plant is regulated at a satisfactory pressure for the later removal of heavy hydrocarbons which would otherwise cause problems in the liquefaction, storage and transport phases of the project. For high pressure gas, i.e. higher than say 55 bar, this may involve a pressure let-down stage. Exceptionally, where low pressure associated gas is involved, for example the Das Island plant, it may be necessary to boost the inlet pressure of the feed gas with compressors.

Liquid hydrocarbons separating out of the feed gas after pressure let-down may be piped either to storage for subsequent sale or direct to the fractionation unit for further treatment.

Metering the quantity of gas supplied to the plant is obviously necessary, especially if the gas supplier(s) is not the same as the owner of the plant which is often the case.

Acid Gas Removal

Any carbon dioxide, hydrogen sulphide and other sulphur compounds present in the feed gas are removed at this stage to levels which are sufficiently low to avoid them freezing out in the liquefaction unit. Typically this means reducing the carbon dioxide content to about 50 ppm (by volume) and sulphur compounds to less than 3 ppm. By the same token, such removal purifies the LNG produced, as well as the NGL which often requires no further treating.

There are various processes available to achieve this mostly involving a solvent which absorbs the impurities and releases them in a regenerator: see Chapter 5 for further details of such processes.

If sufficient hydrogen sulphide is present, a sulphur recovery unit may be installed to convert sulphur compounds into elemental sulphur, a saleable product.

Mercury Removal

In some instances natural gas can contain very small amounts of mercury, i.e. up to say 300 micrograms per normal cubic metre of gas (Nm^3), exceptionally even higher, usually in the form of elemental mercury vapour. All this mercury has to be removed,

Figure 6.2 **North West Shelf LNG Plant, Australia**

The world's largest mobile crane lifting a 380 tonne Sulfinol absorber column into place. Courtesy of Woodside Offshore Petroleum Pty. Ltd.

which means in practice down to about 0.1 microgram/Nm^3, the lowest detectable concentration using current technology, to prevent corrosion of pipework and equipment made of aluminium or aluminium alloys.

Mercury removal can be achieved in several ways, for example, by passing the feed gas through sulphur-impregnated carbon, whereby the mercury in the gas reacts (chemisorption) with the sulphur to form a stable mercuric sulphide, or by adsorption onto macroporous alumina; there are other adsorbents and processes. As these adsorbents are not regenerable, disposal of the spent material, which can contain up to several per cent mercury by weight, has to be very carefully controlled for safety and environmental reasons.

Location of the mercury removal unit in the process sequence will be influenced by such considerations as the composition of the feed gas, the materials used in the various units of the plant, and the amount of water vapour present; the efficiency of

Figure 6.3 **Malaysia LNG Plant: Bintulu, Sarawak**

Courtesy of Malaysia LNG Sdn. Bhd.

the mercury adsorber is reduced at high water vapour percentages. In the Arun plant it is located before the acid gas removal unit, in some other plants after the dehydration unit.

Apart from any mercury that may be present in the feed gas, care has to be exercised that there are no sources, e.g. mercury-containing instruments, that could introduce extraneous mercury into the plant's systems.

Dehydration

The treated sweet gas leaving the acid gas removal unit is saturated with water vapour as the sweetening solvents are generally aqueous solutions. Before cooling the gas below zero degrees centigrade it must be dried in order to avoid icing up in the heavy hydrocarbon separation unit, liquefaction units, storage and ships. This is usually done in two stages.

First, the bulk of the water is condensed and separated from the gas stream by cooling. Cooling is limited to temperatures above the point at which water would form hydrates with hydrocarbons.

Second, the remaining water vapour is removed down to less than 1 ppm usually by adsorption on molecular sieves. At least two of these driers are required for each train; one in service while the other is being regenerated by a heated stream of dry gas.

Heavy Hydrocarbon Separation

The design of this unit will depend on the quantities of ethane and other heavier hydrocarbons present in the natural gas, by the need to extract any heavy hydrocarbons that would otherwise freeze out in the very low temperatures of the liquefaction unit, and by the need to meet the LNG specifications as agreed with the LNG buyer. In addition to these considerations, sufficient hydrocarbon compounds – methane, ethane, propane and butane – will need to be extracted to provide refrigerant components for the liquefaction cycle, while for rich gas streams the removal of hydrocarbons heavier than methane for separate sale is a business opportunity in its own right.

Depending on the composition of the gas stream, there are a number of ways to recover these heavier hydrocarbons of which the most common is cryogenic scrubbing. Typically this involves cooling the natural gas against boiling propane down to about $-30^{\circ}C$ in one or two steps. The liquid mix thus knocked out is then piped to the fractionation unit and the stripped natural gas, predominantly methane and ethane, to the liquefaction unit for further cooling.

However, if the feed gas pressure is near its critical value, then a simple cryogenic scrub cannot be used to remove heavier hydrocarbons. In such cases an expander cycle or an oil scrub is used.

The expander cycle involves an expander-compressor combination in which the natural gas is cooled by expansion, thus taken away from its critical pressure to allow separation of the liquid phase. The compressor, driven by the expander, restores the pressure of the stripped gas.

Oil scrubbing uses a heavy oil to raise the critical pressure of the system thereby allowing the process to be operated at the high inlet pressure of the feed gas. Oil scrubbing may be less economic than the expander cycle when high recoveries of liquids are required or when a light feed gas is being treated.

Figure 6.4 **Two LNG Trains under Construction: North West Shelf Project**

Courtesy of Woodside Offshore Petroleum Pty. Ltd.

Fractionation

The liquid mix from the heavy hydrocarbon separation unit is fed typically to a succession of distillation columns – the de-methaniser, de-ethaniser, de-propaniser and de-butaniser – each stripping out one component from its top end. The bottom stream of each column is fed into the next one so that from the last column flows a de-butanised stream of pentanes and heavier hydrocarbons, sometimes referred to as 'gasoline' or more usually as natural gas liquids (NGL).

These pentanes and heavier hydrocarbons can be either spiked into any heavier stock available, e.g. crude oil, or stored and then shipped out in conventional oil product tankers. In the latter case, the critical specification is vapour pressure, which requires tight operation of the de-butaniser, or even splitting out of the lighter pentane.

Part of the streams of methane, ethane, propane and butane from the tops of the columns are used preferentially to replenish the stocks of mixed refrigerant components. However, the bulk of these streams is either re-injected into the liquefaction unit or disposed of separately. In some cases they may not meet commercial purity specifications and will then require further treating before disposal.

NGL Treatment
If despite the acid gas removal phase the products of the fractionation unit do not satisfy market specifications, they will need to be 'sweetened' in an NGL treatment unit to remove residual hydrogen sulphide, carbonyl sulphide (COS) and mercaptans which may have become concentrated in these products.

COS can be removed by washing using one or more of a great many solvents. When absorbed into aqueous solutions, it combines with water and partially hydrolyzes to carbon dioxide and hydrogen sulphide. With some solvents COS reacts to form unstable compounds and, with others, stable compounds which permanently degrade the solvents.

Traces of mercaptans and hydrogen sulphide are removed by adsorption on beds of molecular sieves.

Shell's Sulfinol and ADIP processes are two of the well-established sweetening processes and are described in Chapter 5.

Liquefaction
There are three main processes currently in use in liquefaction cycles of base-load plants. They are:

- *the pure refrigerant cascade process*

- *the mixed refrigerant process*

- *a combination of the pure refrigerant cascade process and the mixed refrigerant cycle, i.e. the pre-cooled mixed refrigerant process.*

The compression power requirement of a liquefaction cycle is one of the major factors to be considered when evaluating alternative processes. This is because refrigerant compression absorbs at least two-thirds of the total fuel consumed in the plant.

The efficiency of a liquefaction process is frequently expressed as the refrigerant compression power required per unit of LNG produced.

Pure Refrigerant Cascade Process
This process consists of a combination of three cooling systems each comprising a compressor, a condenser, an expansion valve and an evaporator (or heat

exchanger). Typically, propane, ethylene and methane are used in three cooling cycles to provide cooling at progressively lower temperatures.

In the first cycle, the propane refrigerant is condensed at high pressure by cooling water (or air). Then the pressure of the liquid propane is let-down through the expansion valve, and it can vaporise at a lower temperature by condensing the ethylene of the second cycle, as well as cooling the natural gas down to some $-30°C$,

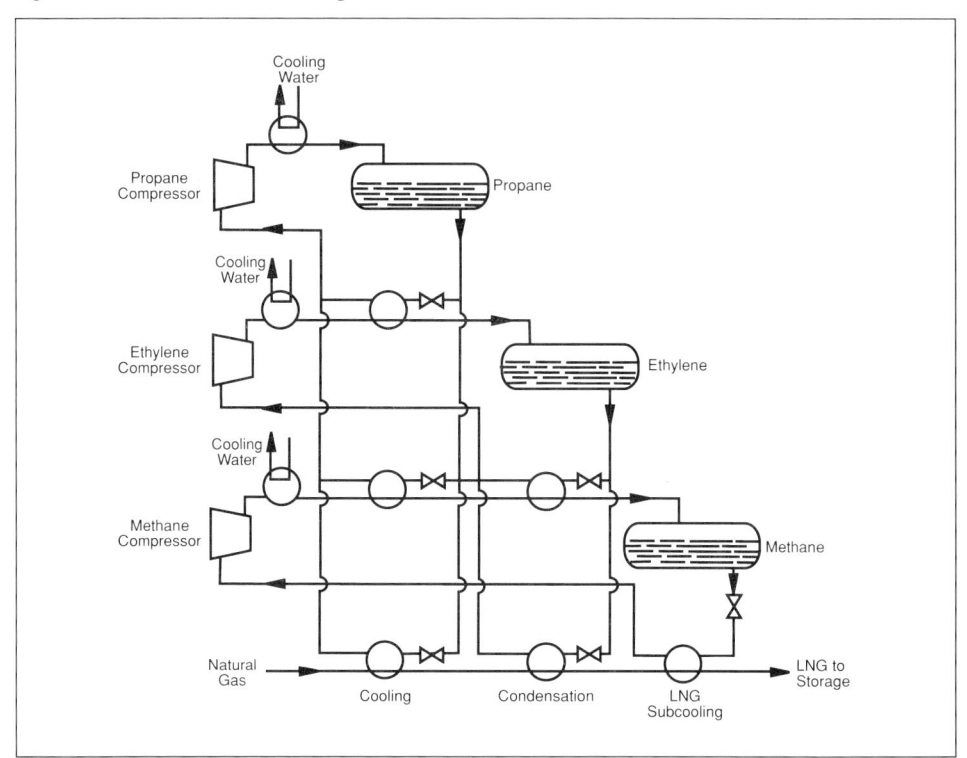

Figure 6.5 **Schematic Pure Refrigerant Cascade Process**

all in a series of evaporators. Finally, the propane vapours are recompressed to the initial high pressure at which they can be condensed at ambient temperature.

Cycles two and three work on the same pattern. Ethylene is condensed under pressure by the propane of cycle one and, once depressured, vaporises by cooling the natural gas down to about $-100°C$. In the last cycle, methane is condensed by the ethylene and, after pressure let-down, is allowed to vaporise by cooling the natural gas to its complete liquefaction temperature of about $-160°C$.

As an efficiency device, when there is an outlet for low pressure fuel gas in the plant, end flash should be considered: cooling of natural gas in the methane evaporator is left incomplete at, say, $-150°C$ and a pressure of 5 bar, thus saving compression power. The last step is done by expanding to atmosphere pressure, thus

Figure 6.6 **Schematic Mixed Refrigerant Process**

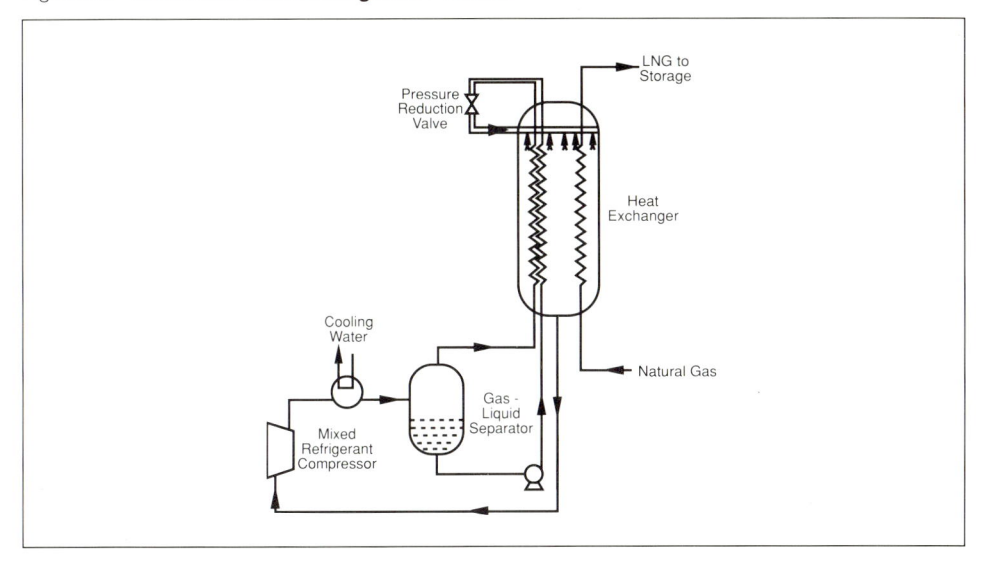

letting part of it (say 5 per cent) revaporise. That flashing brings the liquid at the atmospheric equilibrium temperature of $-160°C$.

End flash also rids the LNG of any excess nitrogen which being lighter flashes preferentially. It can be a decisive factor in considering end flash.

The advantage of the pure refrigerant cascade process is that provided each refrigerant is arranged to cool in several stages, a high efficiency can be achieved. The disadvantages are that complex refrigerant piping and numerous evaporators result in high capital costs and that the ethylene refrigerant losses have to be made up from imported ethylene or by on-site manufacture.

A schematic version of this process is shown in Figure 6.5. This is simplified in that each evaporator is shown as a single unit whereas in practice it is a combination of units at decreasing pressure levels corresponding to different temperatures, each unit being connected to a different suction of the refrigerant compressor.

This type of process is employed in the Arzew GL4-Z and Kenai plants.

Mixed Refrigerant Process

Whereas with pure refrigerants a series of separate cycles are involved, with mixed refrigerants – methane, ethane, propane and nitrogen, sometimes butane and pentane as well – condensation and evaporation take place in one cycle over a wide temperature range down to around $-160°C$.

After compression, the mixed refrigerant is partially condensed against cooling water and sent to a gas-liquid separator. The liquid and vapour are distributed over the tubes in the heat exchanger and are condensed completely. After pressure reduction, gradual evaporation provides refrigeration to liquefy the natural gas.

The process is extremely simple – see Figure 6.6 – but power consumption is substantially greater than for the cascade process. Examples of this process are the Skikda GL2-K and GL3-K plants, Algeria.

Better efficiencies can be achieved with the liquid providing refrigerant at an intermediate level in the lower part of the main heat exchanger to cool the natural gas and partially condense the vapour refrigerant, instead of recombining the vapour and liquid in one stream. This can be repeated in many steps to provide sufficient refrigeration to produce LNG. The Libya plant uses this stepwise process.

Another variant is the two-pressure mixed refrigerant cycle which has two separate exchangers operating at two different pressures. The Skikda GL1-K plant is based on this process.

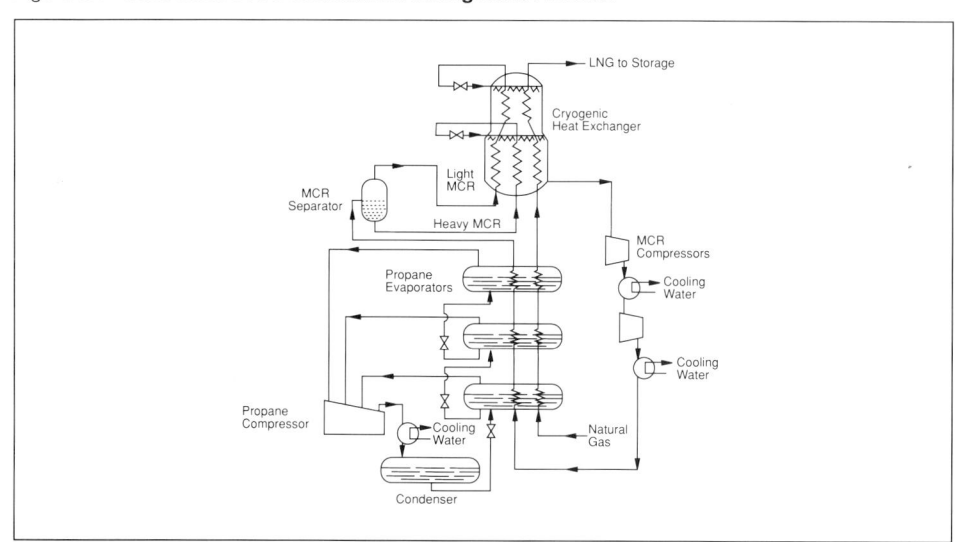

Figure 6.7 **Schematic Pre-cooled Mixed Refrigerant Process**

Pre-cooled Mixed Refrigerant Process

This is a combination of the pure refrigerant cascade process and mixed refrigerant cycle and was first developed for the Brunei liquefaction plant. This process also merges several discrete cycles into one cycle, and although propane pre-cooling is added, this is well justified by increased efficiency.

Propane is first condensed by cooling water and then with heat exchangers in three or four stages, cooling the feed gas stream and mixed refrigerant to about $-30°C$. After compression, the mixed refrigerant is likewise first cooled by water and then, as already indicated, by propane. The mixed refrigerant is Air Products & Chemicals Inc's proprietary Multi-Component Refrigerant (MCR) which, according to one published report, consists approximately of 5 per cent nitrogen, 10 per cent propane, 40 per cent ethane and 45 per cent methane.

Figure 6.8 **Air Coolers of one Train: North West Shelf LNG Project**

Courtesy of Woodside Offshore Petroleum Pty. Ltd.

At this stage, and before the MCR is fed into the cryogenic heat exchanger, it is separated into two fractions, a light MCR and a heavy MCR. Natural gas, already cooled to around $-30°C$, is fed into the bottom of the cryogenic heat exchanger with both MCR fractions and distributed through spirally-wound tube bundles.

The heavy MCR fraction leaves the top of the lower part of the heat exchanger and, after pressure let-down, is separated into vapour and liquid at a temperature of about $-110°C$. They are then reintroduced into the shell of the heat exchanger. The liquid is sprayed over the tube bundles while the vapour mixes with the vapour and liquid flowing downwards from the shell of the upper bundle.

In the top part of the heat exchanger, the light MCR fractions and natural gas are cooled to around $-160°C$ by spraying light MCR after pressure let-down over the remaining bundles. At this point the natural gas is liquefied and fed to storage. Low pressure MCR vapours are collected at the bottom of the heat exchanger, compressed and recycled. The whole process is shown schematically in Figure 6.7.

Apart from Brunei, this basic process, or variants thereof, is employed in the Abu Dhabi, Indonesia (both Arun and Badak), Malaysia and Australia plants and in two of Algeria's liquefaction plants. The Australia North West Shelf project uses air cooling instead of water cooling, and G.E. Frame-5 double-shafted gas turbine drivers instead of steam turbines for compression and power generation. Some other projects, e.g. Arun, also use gas turbines as well as steam turbines. In fact all MCR processes are of the same family, each one representing some improvement over the previous version. A further refinement would be to replace the propane pre-cooling by a primary MCR cycle ending up with a double MCR process.

These sections have only outlined the three basic liquefaction processes in current use. Each and every plant has its own special features and design characteristics and, of course, technology does not stand still. Inevitably, state of the art plants are more efficient and sophisticated than their predecessors, but the basic concepts are fundamentally similar.

Storage and Loading Facilities

As LNG storage for liquefaction plants involves the same essential design concepts as storage for receiving terminals and peak shaving plants, for convenience sake they are discussed collectively in Chapter 7. However, it is appropriate to point out that assessing the storage capacity for a liquefaction plant is usually a much simpler task than that for a large receiving terminal which in addition to the receipt of LNG, sometimes from different sources, may also have to provide capacity to handle local market seasonal and peak load demands.

Similarly, and to avoid repetition, as loading LNG into ships involves much the same procedures and systems as unloading at receiving terminals, the reader should refer to Chapter 9 for details on this subject.

Utilities

The utilities required for an LNG plant are large because of the power and cooling water needs of the liquefaction process. A modern plant, generating its own power and utilities requirements, consumes about 10 per cent of its feed gas (disregarding the inert components) as fuel. That huge ratio, by comparison to petrochemical plants and refineries, reflects the cost of cold generation.

The use of that fuel gas, and consequently the breakdown over the various utilities, is determined by two factors: the driving medium for the refrigerant compressors – steam turbines versus gas turbines – and the cooling medium – water versus air – the main duty of which is the condensation of the refrigerant. Gas turbines are the modern option that is superseding steam turbines, but the cooling medium remains an open question depending upon the site conditions.

Cooling water can be either a circulating fresh water system with cooling towers if a suitable nearby source is available, or predominantly a once-through seawater system.

Nitrogen is also required as a refrigerant component and as an inert gas to purge storage tanks, pipelines, ships' tanks, etc., as necessary. Air separation units are provided in conjunction with liquid nitrogen storage so that nitrogen is available as and when required.

Finally, a fuel gas system is needed to fire steam boilers (or gas turbines) and process furnaces. Supplies of fuel gas can comprise a mix of boil off gas from storage and loading facilities, gas from process units, flash gas and feed gas to the plant itself.

Safety

Although this is the last section of this chapter, in practice it will be among the first of the major considerations affecting where the plant should be sited, its layout, the processes, type of storage, materials and equipment to be used and where new infrastructure will be located. Critical to these considerations is the necessity right from the outset to ensure the safe arrival, berthing, loading and departure of the ships that will service the plant.

In all these and related matters plant designers and engineers work very closely with marine experts, local and national authorities and any other specialists who can contribute to finding the optimum, practical, reliable and safe solution.

Once these various decisions have been taken, safety and emergency procedures will be drawn up, personnel trained and regular safety audits undertaken covering all aspects and phases of construction, start up, operation and maintenance of the plant right through its operating life. Some of these matters are elaborated on elsewhere in this book.

Chapter 7

LNG AND OTHER TYPES OF STORAGE

Natural gas can be stored in liquefied form in above-ground or in-ground tanks, or in its gaseous state under pressure in suitable geological formations, including depleted oil and gas reservoirs, or man-made caverns. Underground gaseous storage is utilised mainly for seasonal load balancing purposes, whereas LNG storage can be used in a continuous operational mode or partly or wholly for load balancing and peak shaving. Load balancing is essentially evening out periods of relatively low and high demand to improve the overall load factor (for definitions of load factor see Chapter 19) of a gas supply system the importance of which is discussed in Chapter 10. Peak shaving, as the name implies, is the supply of gas during very high levels, or peaks, in demand which may occur during a few days or weeks of a year when ambient temperatures are very low. Typical ways and means of load balancing and peak shaving are illustrated in Figure 7.8.

Figure 7.1 **Cross Section of a Typical Above-ground LNG Tank**

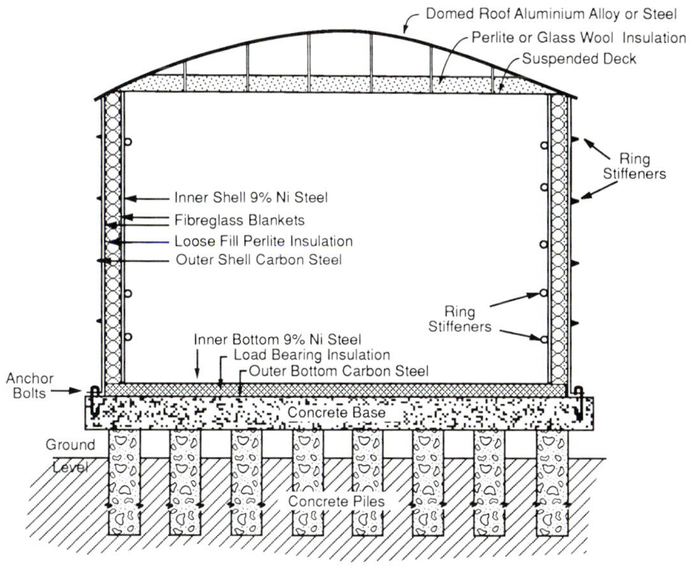

N.B. Not shown are the loading/unloading pipework and pumps, measuring equipment, inspection manhole, surge and splash plates, etc.

Apart from one in-ground 'frozen hole' storage unit of 38000 m3 capacity built at Arzew, Algeria, in 1966, all LNG storage tanks at existing liquefaction plants are of the above-ground type (including those of that type that are semi-buried) as is all storage at existing European and North American LNG reception terminals. Another exception used to be Canvey Island, England, where four frozen hole units (each of 21000 tonnes capacity) were built in 1969 but these were decommissioned in 1983 when they were no longer required.

In Japan, both above-ground and in-ground LNG tanks are employed at reception terminals, although in more recent years the latter have generally been preferred. This is because they can be spaced closer together, an important factor where land for terminals is very limited, to be more in harmony with the environment and may be less affected should severe earthquakes occur. However, this is not to suggest that above-ground tanks are in anyway less secure or reliable. Indeed their performance and safety record over the last 30 years or so has been exemplary.

As there are many variations in the detailed design aspects of LNG tanks in service today, the descriptions given in this chapter are confined to basic concepts and therefore do not cover all the many designs that are available. Subject to this qualification, the concepts described apply equally to LNG tanks for base-load liquefaction plants, LNG receiving terminals and LNG peak shaving plants.

LNG Above-ground (and Semi-buried) Tanks

There are approximately 250 above-ground LNG tanks currently in service, the majority of which are of the double-wall metal tank configuration. Capacities range up to 100000 m^3 (liquid). The principal alternative is the prestressed concrete tank. There are a number of different designs and in several instances tanks of the prestressed concrete type are semi-buried. Capacities range up to 120000 m^3.

Double-wall metal tanks – see Figure 7.1 – comprise a gas and liquid tight inner tank or shell usually made of 9 per cent nickel steel, or less frequently of aluminium. Aluminium is not normally used for very large capacity tanks because of the thickness of aluminium required which is difficult to weld satisfactorily, and because of the greater coefficients of thermal expansion that must be allowed for in the design of the tank. Stainless steel is also used but this tends to be confined to small capacity tanks for cost reasons.

As the outer tank is not subjected to very low temperatures, it is invariably made of carbon steel for cost reasons. This tank acts as a pressure barrier to eliminate any differential pressures between the inner and outer tanks, as a means of containing the insulation material, and as an additional barrier to water, fire and any objects which may hit the tank accidentally or deliberately in the case of sabotage. However, this outer tank is not designed to contain very cold liquids so the overall concept is of one (inner) tank to contain the LNG.

Fibreglass blankets are placed on the outer surface of the inner tank, and frequently on the inner surface of the outer tank as well. Perlite, which is an excellent insulation material and is inexpensive and fire resistant, is the most commonly used material for filling and insulating the annular space between the two tanks.

The outer tank sits on and is bolted to a concrete base or foundation which is heavy enough to balance maximum uplift when the tank is empty of liquid. This base may be built into the ground, in which case soil heaters are provided, or built on concrete piles leaving an air space between the ground and the bottom of the base which eliminates the need for soil heating. In both cases the avoidance of any settlement is essential to prevent stresses on the tank and the initiation of any cracks in the insulation. Insulation of the tank bottom usually comprises load bearing foam glass blocks.

The domed gas tight roof is made of aluminium alloy or steel and supports a suspended insulated deck which ensures that the roof is kept at or close to ambient temperature. Here again Perlite may be used as the insulation material or glass wool. Loading pipes, unloading pumps, measuring and other equipment to operate the tank are now almost invariably inserted through the roof as this helps to preserve the integrity of the tank structure. Typically, boil off rates for tanks of around 50000 m^3 capacity are less than 0.1 per cent per day.

Earthen or concrete bund walls or dykes of sufficient height and at a sufficient distance from the tank are built capable of containing all the liquid content of a full tank in the unlikely event it should fail. The concern here will be to ensure that the area circumscribed by the walls is not too great, so as to avoid an unnecessary high vaporisation rate should there be a large spill.

Figure 7.2 **LNG Peak-shaving Plant at The Maasvlakte, The Netherlands**

Courtesy of N.V. Nederlandse Gasunie

At some locations, e.g. Gasunie's peak-shaving plant at the Maasvlakte, The Netherlands, a separate concrete container has been built around the double-wall metal tanks to give added protection from possible damage from the outside and to confine the LNG in the event of a major leak of the inner tank. Alternatively, as for example Distrigaz's reception terminal at Zeebrugge, Belgium, the tanks may have to be semi-buried to meet local authority requirements which in this instance were that the height of the tanks should not exceed 30m above ground level. This was achieved by constructing in-ground caissons in which the tanks were then built.

There are various types or designs of above-ground prestressed concrete tanks. Briefly, the principal concepts are:

i) *An outer prestressed concrete wall with an inner 'stand alone' metal tank and with the annular space filled with an insulation material. The roof can be either carbon steel, or prestressed concrete supported by the outer concrete wall, with a suspended insulated inner deck.*

ii) *Inner and outer prestressed concrete walls with insulation filling the annular space. The floor of the inner tank will usually be 9 per cent nickel steel over a load bearing insulation. In some designs the inner concrete wall is also clad with Ni steel. Different types of roofs are used.*

iii) *An outer prestressed wall, with interior wall and floor insulation, and with an inner metal membrane liner for liquid containment. The roof can be either prestressed concrete supported by the outer concrete wall, or carbon steel, with a suspended insulated inner deck.*

iv) *An inner prestressed concrete tank, which is the cryogenic containment barrier, with an outer steel tank for insulation material containment which also supports the metal roof and the suspended insulated deck.*

v) *Double-wall prestressed concrete tanks with a concrete berm in the annulus. The inner prestressed wall and floor have a 9 per cent nickel steel lining which serves as the liquid barrier. The annular space between the inner wall and the berm is filled with insulation material. The outer wall supports a steel or prestressed concrete roof with its suspended insulated deck.*

The advantage of many, but not all, above-ground concrete tank designs is that they can support an earthen berm. This gives the tank higher resistance to external impacts and added protection in the event of a fire. Moreover, with some design concepts the outer concrete tank and its berm are capable of withstanding the asymmetrical hydro-dynamic impact that would be the consequence of a sudden failure of the inner tank. If these conditions are satisfied under any conceivable mode of failure of the inner tank, then the concept is known as 'double integrity'.

An example of the double integrity tank concept is the tankage at the Bintulu, Malaysia, liquefaction plant – see Figure 7.3 – where from the inside outwards the

65000 m^3 capacity tanks comprise a 9 per cent nickel steel shell, a fibreglass blanket, Perlite, polyurethane foam coated carbon steel lining and a re-inforced concrete wall. Each tank is surrounded by an earthen berm tapering to the top of the concrete wall from ground level. The floor comprises nine per cent nickel steel, foam glass blocks, a carbon steel lining and a re-inforced concrete base with soil heaters underneath. Another example is the tanks of the Australian North West

Figure 7.3 **LNG Storage, Bintulu, Malaysia**

One of the four 65000 m^3 capacity tanks of Malaysia LNG Sdn. Bhd.

Figure 7.4 **LNG Storage, Withnell Bay, Australia**

Four 65000 m^3 capacity double integrity tanks. Courtesy of Woodside Offshore Petroleum Pty. Ltd.

Shelf liquefaction plant which are encased in concrete and are semi-buried – see Figure 7.4.

LNG In-ground Tanks

Selection of tank design and the method of construction will depend upon satisfying local regulations and national and/or international codes of practice, capital and operating cost considerations and, particularly in the case of in-ground tanks, the

Figure 7.5 **Cross Section of a Typical In-ground LNG Tank**

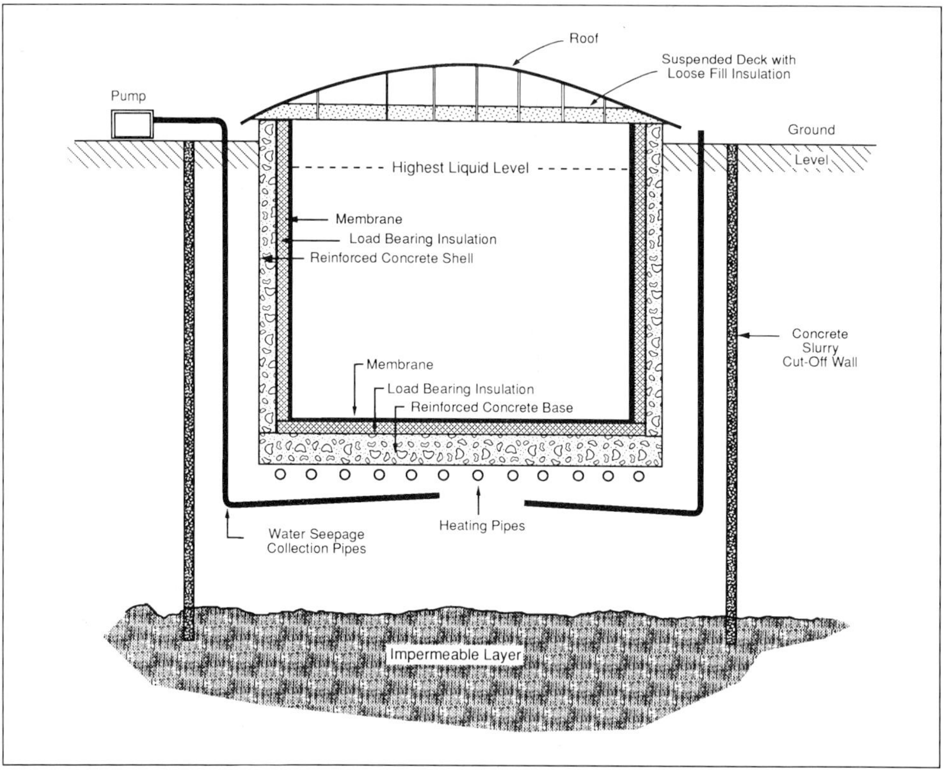

ground conditions of the site. A fundamental requirement is that the ground of the site is stabilised so that the tank will not be affected adversely by any future settlement of the ground.

The principal problem encountered during construction and subsequent operation of an in-ground tank is dealing with underground water. Thorough investigation of the underground water level, water pressure and water permeability of the ground is therefore essential. If soil freezing is allowed to develop unchecked once the tank is in service, the phenomenon of frost heave can occur which can affect the integrity of the tank. However, frozen soil as such around the tank can help to increase the strength, air-tightness and safety of the structure.

There are various methods of stopping water entering the site during excavation and of preventing the problems associated with frost heave once the tank is built and in service. These include installing a subsurface diaphragm or slurry cut-off wall around the site of the tank down to an impermeable soil layer, caisson type construction, providing an independent bottom concrete slab of sufficient thickness to withstand the hydrostatic and frost heave pressures acting on it, surrounding the tank sides and bottom with seepage collection pipes to drain off underground water, etc. Soil and water conditions and the size of the tank will tend to dictate which methods are most appropriate for the tank design and site in question, but in all cases soil heating systems are virtually indispensable.

The actual structure of the tank, starting from the inside, usually comprises a thin (about 2 mm) liquid and gas tight stainless steel membrane, an insulation material for heat insulation, and a shell of reinforced concrete for supporting liquid (LNG), earth and water pressures. For a large tank the shell can be 3m thick. The domed roof is a steel supported shell structure capable of sustaining the pressure of the boil off gas, with a suspended insulated deck – Figure 7.5 shows a typical structure. Filling, unloading and boil off gas pipework, the submerged unloading pump, measuring and other equipment are inserted through gas tight orifices in the roof. The boil-off rate for a large in-ground tank is about 0.1 per cent per day.

While different membrane configurations are used they all perform the same functions of providing a gas and liquid tight container, the mechanism for absorbing thermal expansion and contraction caused by temperature changes, and for transferring liquid and gas pressures through the insulation material to the tank

Figure 7.6 **In-ground LNG Tanks**

Two 130000 m^3 capacity LNG tanks. Tokyo Gas's Sodegaura terminal. Courtesy of Diamond Gas Operation Co. Ltd.

shell, while at the same time having sufficient fatigue strength to withstand repeated pressure and temperature changes.

In addition to its insulation function qualities, the insulation material, usually a rigid polyurethane foam, has to have sufficient compressive strength to transfer the liquid pressure from the membrane to the concrete shell, creep strength, fire and humidity resistance and good workability qualities.

Figure 7.7 **In-ground Storage**

The interior of one of the tanks shown in Figure 7.6. Courtesy of Diamond Gas Operation Co. Ltd.

Soil heating systems can comprise electric cable heaters or the circulation of warm water or a glycol brine mixture through pipes around and underneath the tank.

There are now 50 in-ground tanks in service in Japan ranging from 10000 m^3 (the first built in 1970) to 140000 m^3 (approximately 65000 tonnes of LNG) capacity and larger capacity tanks are being designed. Typically, a tank of 130000-140000 m^3

would have an inner diameter of some 60-65m and a depth of 40-45m, although slurry walls, if installed, can extend down to about 100m. In the latter regard, the main purpose of the slurry wall was to exclude water during excavation, but a more recent development has been to incorporate the slurry wall as an integral and permanent part of the tank side wall which can then be made thinner. This reduces the time and cost of construction.

Generally speaking, in-ground tanks are more difficult and, excluding land costs, more expensive to construct than above-ground tanks. However, they may be closer spaced which saves land cost, are less obtrusive and, as all the LNG stored is below ground, there would be no spillage of LNG at ground level in the event of a tank failure.

Rollover

Rollovers are a result of stratification from filling any type of cryogenic tank with LNG of different densities which remain unmixed, or of autostratification due to the preferential loss of nitrogen if present.

If a tank containing LNG is further filled with LNG of a different density, then it is possible for the two liquids to remain unmixed, forming independent layers. This can occur if the added LNG is denser than the LNG already in the tank (the heel) and is filled at the bottom, or if less dense than the heel and filled at the top.

Rollover, which is the rapid mixing or homogenization of the two stratified layers, occurs when the densities more or less equalise due to changes in the temperature and composition of the two LNG layers brought about by heat absorption and weathering. If the mixed LNG then has temperature and composition such that it is superheated with respect to the vapour pressure in the tank, there is a sharp increase in the vapourisation rate. Under certain conditions, this can trigger the emergency vent system and relief valves of the tank to relieve the excess vapour pressure. If this should exceed the capacity of the system, it could have hazardous consequences.

Stratification can obviously be avoided by filling LNG of different densities into separate tanks. However, if insufficient storage capacity is available to accept separate cargo lots without mixing with LNG already in storage, then stratification can normally be prevented by using jet mixing nozzles, alternating top and bottom filling, filling through multi-orifice tubes, etc. Increasingly, tanks are being fitted with instruments to monitor temperature, density and boil-off to detect stratification so that preventative action can be taken before rollover can occur.

Autostratification can arise if LNG containing more than about one per cent nitrogen is added to a nitrogen-free LNG. Nitrogen as the most volatile component boils-off preferentially causing the saturation temperature of the remaining liquid to increase. In the case of the nitrogen-free LNG the preferential loss of methane, which is the most volatile component, causes increases in both the saturation temperature and density of the remaining liquid. The behaviour of the two liquids

Figure 7.8 **Schematic Load Duration Curve**

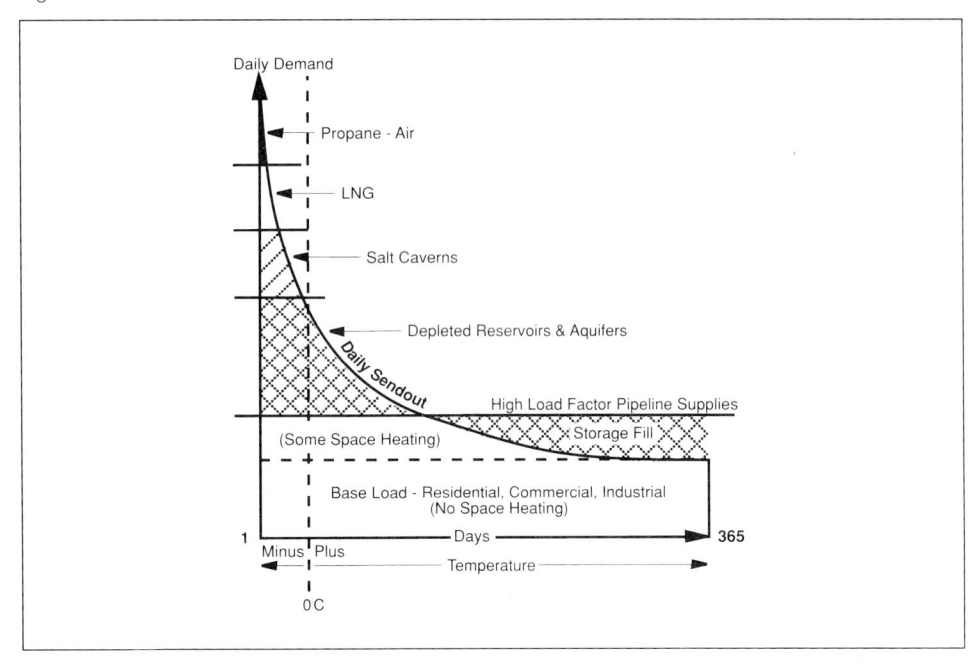

thereafter is similar to a fill-induced stratification except that if after rollover the mixed LNG still contains appreciable nitrogen, the process can repeat. Autostratification is best prevented by ensuring that the nitrogen content of the LNG is kept sufficiently low.

Safety

Since the Cleveland, Ohio, disaster of 1944 when an LNG tank failed, spilt its entire contents which then vapourised and caught fire, there have been no reported major incidents of this nature with any type of LNG storage. The cause of the Cleveland disaster was never precisely determined, but was most probably caused by unsuitable tank shell and/or insulation materials. The industry has learned much since then on the nature and behaviour of both LNG and the methods and materials to handle and contain it safely.

Safety is one of the first considerations in designing and siting an LNG tank farm, and includes the need to protect the lives and health of plant personnel and the general public that may be living and working nearby. While it is not possible to make any operation involving flammable material absolutely safe, it is possible and necessary to anticipate and obviate the main risks.

Various design features to contain any spillage of LNG in the unlikely event of a tank failure have already been discussed in previous sections, while the siting and layout of a tank farm will reflect site-specific circumstances. However, there still remains the risk, however slight, of the formation of a flammable gas cloud that could result from, for example, an accidental spillage of LNG or the release of LNG vapour following the failure of a tank roof or the operation of relief valves.

The main hazard associated with an LNG vapour cloud is that it could ignite, and while detonation of an unconstrained vapour cloud is not possible, deflagration accompanied by some measure of over-pressure cannot be ruled out completely. The risk is reduced if the size of the spillage is reduced or so contained as to minimise the size of the vapour cloud formed.

Apart from tank design and layout considerations, prompt detection of gas, smoke or fire is therefore essential so that action can be taken to minimise the consequences of a spill or vapour release. In addition to gas, smoke and fire detection systems, LNG tank farms will have water spray cooling systems, high expansion foam and dry chemical powder extinguishers, with grading and drainage into sumps where the vapour generated from a spill can be better regulated. Finally, of great importance is the establishment of safe operating procedures and the training of plant personnel in both their normal responsibilities and how to handle emergency situations should they arise.

Seasonal Storage

In markets that experience appreciable seasonal variations in ambient temperatures, and where space heating has become a significant outlet for gas, storage will be needed in or reasonably close to the main areas of consumption to supplement base load supplies during periods of high demand. The principal methods of providing large capacity underground storage for seasonal use are aquifers, depleted oil and gas fields and salt caverns. Disused mines are also employed in North America, and several rock cavern projects are being developed in northern Europe.

Seasonal storage is filled progressively during periods of low demand which helps to improve the load factor of the pipeline system which supplies gas for storage. Part of the injected gas, which can be as much as 50 per cent, remains in the storage as 'cushion gas' for operational purposes, and in some cases part of the injected gas may be lost by migration out of the formation in which the gas is stored.

While most underground storages are used primarily for seasonal load balancing purposes for periods of up to several weeks or months of the year – see Figure 7.8 – some storages may be employed partly or wholly for peak shaving purposes during very cold days of high demand to augment LNG and propane-air supplies. To perform this latter role the storage facility must obviously have a high daily withdrawal (send-out) capacity; salt caverns are usually inherently more suitable than aquifers or depleted reservoirs for peak shaving.

Apart from its seasonal and/or peak shaving role, another important function of underground storage can be to provide strategic storage. This may be a legal requirement in some countries and in any event will be an important consideration for those markets which rely to a large extent on imported supplies.

Optimisation of the storage facility can include increasing, if practicable, the pore volume of the reservoir layers, e.g. by hydraulic fracturing in a carbonate reservoir, improving the working/cushion gas ratio within the pressure limits set by the gas tightness of the reservoir, optimising the flowrates of production and injection wells, increasing the maximum pressure and reducing the minimum pressure, limiting water production, etc.

Economic considerations will reflect, inter alia, the capital investment required and operating costs involved which will be site specific, the expected duty of the storage, the extent to which gas treatment facilities have to be provided, comparison with such alternatives as may be available, etc. An important element will be working

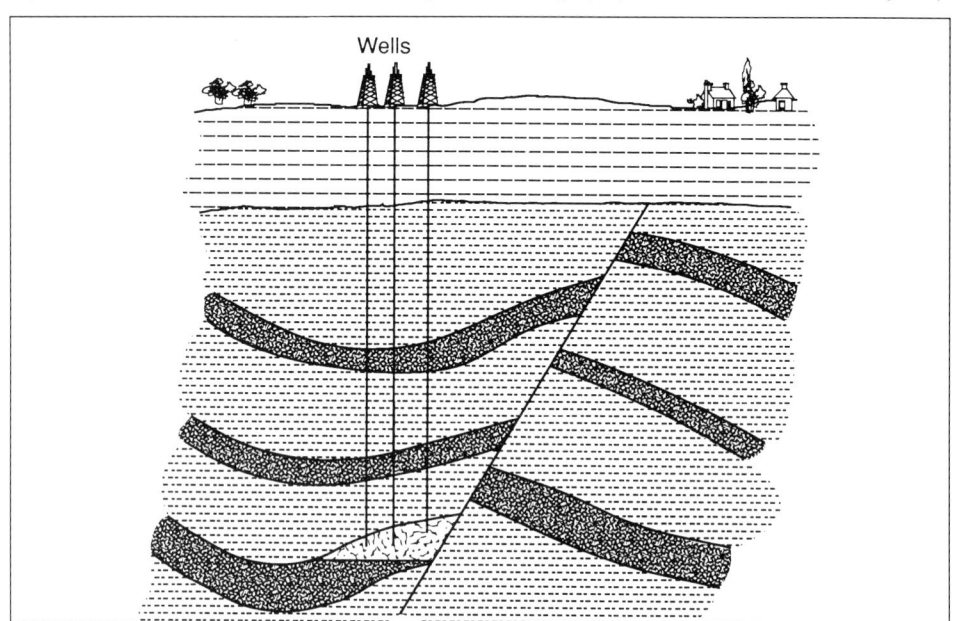

Figure 7.9 **Schematic Structure for Underground Storage (Depleted Gas Reservoir or Aquifer)**

capital for the purchase of gas which may not be sold until some months later, possibly at a different price, and the cost of cushion gas which can account for one-third to one-half of the total investment cost.

As a generality, and there are exceptions, it is more economical to use a depleted gas reservoir than a depleted oil reservoir or an aquifer.

According to a recent Cedigaz report, the first underground gas storage in a depleted gas reservoir was commissioned in the United States in 1916 and is still in service today. By 1990 there were 542 underground gas facilities in the world of which 423 use a depleted reservoir, 82 are aquifers, 33 are salt caverns and 4 are disused

Figure 7.10 **Underground Storage at Bierwang, Germany**

Courtesy of Ruhrgas A.G.

mines. In the United States, the working gas underground storage volume is about 130 billion (10^9) cubic metres (Bcm) with a daily withdrawal capacity rate of 1.2 Bcm, while in West Europe the equivalent figures are some 30 Bcm and 0.6 Bcm at the maximum daily rate. Data for East Europe and the Soviet Union are less precise but certainly as regards both working gas volume and withdrawal rates they exceed in total those of West Europe. Apart from Australia, underground storage is not yet employed elsewhere, although projects of this type are being planned in a number of countries in Asia and Latin America.

Aquifers and Depleted Fields

Aquifers, depleted gas fields and, to a lesser extent, depleted oil fields are the most common means of storing natural gas in any quantity. As their characteristics are not very different and the technology is similar, they are discussed together for convenience sake.

The suitability of an aquifer or a depleted field for gas storage will depend on such features as:

- *being sited reasonably close to the market it will serve;*
- *the geological tightness of the reservoir to prevent excessive leakage and pressure loss;*
- *good reservoir porosity and permeability qualities;*
- *sufficient depth below the surface to allow for safe pressure operation;*
- *easily controllable water condition in the reservoir if water is present.*

Reservoirs may be single or multi-layer formations usually in sands, sandstones and carbonates at depths below the surface of about 500 to 2000 metres. Preferably they should be a thick vertical formation rather than a thin horizontal one.

Many depleted gas fields are too large for effective storage operations as they require too great a volume of cushion gas to increase reservoir pressures to a satisfactory level. Depleted oil reservoirs are not always satisfactory if residue oil is produced in sufficient quantity to interfere with gas withdrawal operations.

Figure 7.11 **Wellhead at Bierwang**

Courtesy of Ruhrgas A.G.

Depending upon the permeabilities of the reservoir layers, production wells which may be open hole, cased and perforated or gravel pack, are drilled typically up to 400 metres apart with production tubings usually between 3 and 7 inches diameter. In all cases compression is used to inject the gas into the reservoir and in some cases for withdrawal as well. Gas saturation in the reservoir of 50 to 75 per cent is common, exceptionally 80 to 90 per cent. Storage capacities, injection and withdrawal flow rates embrace a very wide range. An example, not necessarily typical, is Ruhrgas's

Figure 7.12 **Making Salt Caverns for Gas Storage**

Water reservoirs used in conjunction with leaching out underground salt deposits, near Berre, France.

storage at Bierwang, Germany. This is a depleted gasfield, 1550m below the surface with a working capacity of 1.3 Bcm at a pressure of 160 bar and a maximum withdrawal rate of 800000 m^3/hr. In some storages, e.g. in France, part of the cushion gas has been substituted with an inert gas.

Gas after it is withdrawn will usually require dehydration and, in some cases, desulphurisation and other treatment, before it can be distributed.

Salt Caverns

Salt beds (and salt plugs) of sufficient thickness and of suitable geomechanical properties can be leached out to create underground caverns for gas storage. Such beds are to be found in Zechstein (Permian), Devonian, Sannoisian (stage of Oligocene) and other formations, typically at depths below the surface of 200 to 2000 metres and in thickness of several hundred or even several thousand metres.

Salt is leached with fresh water, with gas oil, or sometimes LPG, being injected into the annular space as a blanket during the process. Brine is discharged from the cavern by displacement with the gas to be stored; rates of displacement are usually around 100 m^3/hr. If local opportunities exist, the brine leached out may be used by the chemical industry, or alternatively deposited in old mines, injected into absorbent deep formations, or run into surface waters provided this is environmentally acceptable.

Cavern depths and dimensions obviously vary with location, but are typically 100 to 400 metres in height, up to 100 metres in diameter, at some 500 to 2000 metres below the surface. Gas capacities, operating pressures and withdrawal rates vary

Figure 7.13 **Compressors for Gas Injection**

Courtesy of Ruhrgas A.G. Four reciprocating compressors for injecting gas into underground salt cavern storage at Krummhorn, Germany. There are three caverns at 1500-1800m below the surface. Working capacity 120 million m^3. Maximum production rate 500000 m^3/hr.

considerably; cushion gas, as distinct from working gas, is normally between a quarter and one-half of the cavern's capacity.

While the composition of the stored gas does not change, its moisture content increases over time during the storage process up to water vapour saturation. Thus all salt caverns are equipped at the surface with dehydration facilities, most of which are operated with triethylene glycol as the absorbent, to remove entrained water before the gas is fed into the pipeline system for sale.

One potential problem with salt caverns is the risk of hydrate formation which can block valves and outlet pipework. Under certain conditions hydrates can form with natural gas in the presence of water vapour at high pressure and low temperature. These conditions can arise at high gas withdrawal rates because the cooling of the gas due to pressure reduction – the Joule-Thomson effect, see Chapter 4 – is considerably greater than the counteracting heat influx from the surrounding salt formation. Hydrate formation can be controlled/prevented by the injection of glycol and/or 'warm' gas at the wellhead.

The world's first salt cavern storage for natural gas was commissioned in Michigan in 1961. Of the 33 now in service, half are in the United States.

Rock Caverns

In a number of countries, e.g. Finland, Sweden, etc., conventional methods of storing natural gas underground are not possible because of unsuitable geological conditions. An alternative is high pressure cavern storage in basement rock, either lined or unlined. These have been used successfully for LPG storage, and for non-pressurised oil products, for some years and are now being considered for natural gas.

Lined rock caverns, in the form of vertical cylinders, are excavated in diameters up to 50 metres and in height up to 100 metres some 150 to 200 metres below the surface. The cavern walls are lined with steel or plastic to achieve gas tightness and supported by concrete walls. Tunnels and shafts are used for transporting the spoil to the surface for disposal and for water draining. Wells for injecting and withdrawing storage gas are drilled in accordance with standard oil and gas industry practice.

The advantage of lined caverns is that the stored gas is never in contact with water and thus gas withdrawn from storage does not require any further treatment before it is distributed. A potential problem is that high gas pressures will initiate cracks in the concrete walls and rock and impose loads on the lining. Ongoing work includes assessing the behaviour of lining materials under these conditions and the possible consequences of gas breakthrough if the lining failed.

Unlined caverns which are excavated in stable bed rock at depths down to 1000 metres, i.e. deeper than lined caverns, achieve gas tightness when all the fractures in the rock surrounding the cavern are filled with water and when this water is always at a greater pressure than the stored gas. Sometimes it may be necessary to inject water to achieve the necessary pressure level.

While unlined caverns avoid the cost of a lining and concrete walls, they need systems to handle and treat water and facilities to dry the stored gas before it is distributed. Hydrate formation may also be a problem.

Rock cavern storage, both lined and unlined, of natural gas is planned to be in service during the 1990s in Finland and may be in Sweden as well. The technical and economic feasibility of storing LNG in caverns is also under investigation.

Chapter 8

LNG SHIP DESIGNS, OPERATION AND OWNERSHIP ASPECTS

Refrigerated Ships: General Aspects

Although several types of ships have been developed over the years for carrying liquefied gases – butanes, propane, ethane, ethylene and natural gas – ranging from fully-pressurised through semi-pressurised to fully refrigerated systems, only the latter type has been used in commercial service to date for the ocean transport of natural gas in liquefied form. Discussion in this chapter is therefore limited to fully refrigerated ships, although it is possible that other systems may be introduced for natural gas transport in years to come. In 1991, there were 65 fully refrigerated LNG ships in service, a further 20 on order/under construction and about another 10 close to the firm order stage. Others will undoubtedly be ordered as new LNG projects are implemented.

Several LNG ships built in the 1960s/1970s were subsequently scrapped or converted into ore carriers when the prospects of finding employment for them did not appear very promising, or because the project they were scheduled to service was aborted. The wheel has since turned a full circle with shipping capacity now in short supply.

Individual cargo capacities of the current LNG fleet range from 25500 m^3 (CINDERELLA delivered in 1965) to 136400 m^3 (EKA PUTRA 1989). However, the majority, some 60 per cent, are around 125000 to 130000 m^3 which over the years has become the preferred size of ships servicing many international base load LNG projects, although capacities are tending to edge up with four ships of 137500 m^3 now on order for the Abu Dhabi-Japan project expansion. Capacities quoted are, of course, in terms of liquid cubic metres.

Because of the low pressure employed – essentially atmospheric – the cargo tanks of these ships do not need to be of pressure vessel form so that although spherical and cylindrical tank forms are employed in some designs, the remaining designs are based on rectangular or trapezoidal flat walled tanks tailored to fit the available ships' hold spaces. The tanks are insulated, and unlike LPG and ethane/ethylene ships, LNG ships do not normally have reliquefaction plants on board. Although this is technically feasible, the complexity and economics involved have not to date justified the installation of such equipment.

A ship designed for LNG may also be able to carry ethane/ethylene or LPG provided this is recognised in the preliminary ship design so that the effects of the higher density cargoes on tank design and support systems are taken into account. However, as such a dual purpose carrier must be designed for the most demanding cargo in terms of material/temperature requirements, it is not normally economic to use a high grade ship for a lower grade cargo on a continuous basis. A shipowner would normally only use a LNG carrier to carry LPG if no LNG cargoes were available.

Design Features : Low Temperature Effects

Figure 8.1 shows the basic construction features of a fully refrigerated LNG ship. The two principal distinguishing features of this design are its insulation and cargo container.

Figure 8.1 **Principles of LNG Ship Construction**

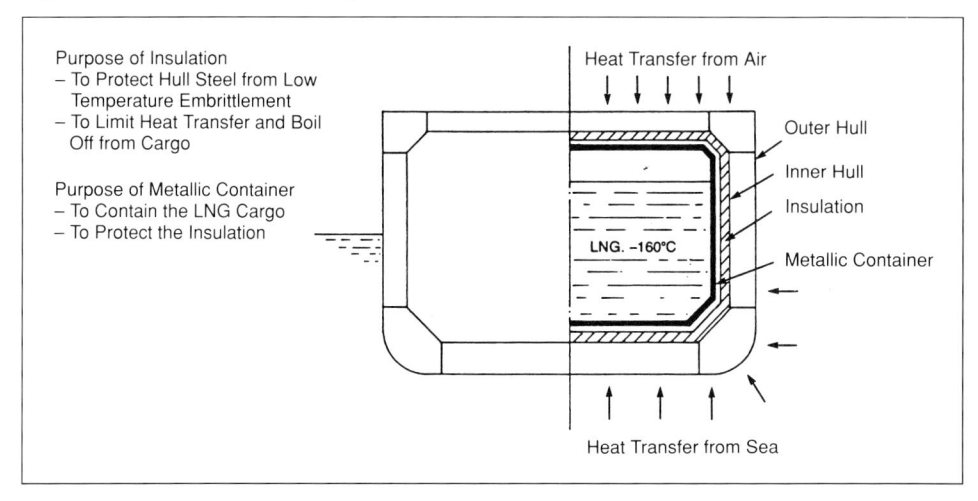

Purpose of Insulation
– To Protect Hull Steel from Low Temperature Embrittlement
– To Limit Heat Transfer and Boil Off from Cargo

Purpose of Metallic Container
– To Contain the LNG Cargo
– To Protect the Insulation

Heat Transfer from Air

Outer Hull
Inner Hull
Insulation
LNG. –160°C
Metallic Container

Heat Transfer from Sea

The mild steel used for the hull of a ship is not suitable for operating at cargo temperatures, because mild steel becomes brittle and fractures at low stress levels if the steel is subjected to temperatures significantly below 0°C. An insulation system is therefore required to prevent steel temperatures falling below an acceptable and safe level. In addition, as the cargo is at its boiling point any heat flow from the outside into the cargo will cause the cargo to evaporate or "boil off". The insulation system therefore serves the dual purpose of protecting the hull steel from low temperatures and reducing the "boil off" of cargo on voyage to acceptable limits.

Most insulation materials are not suitable for direct contact with the cargo as they are not liquid tight. To contain the cargo and protect the insulation requires a metallic container fabricated from a suitable material which maintains the necessary qualities at low (cryogenic) temperature.

Unlike many conventional oil tankers where the cargo tanks may also be used for sea water ballast on the return voyage, the cargo tanks on liquefied gas carriers are used exclusively for cargo. To accommodate water ballast when required, and also to give added protection to the cargo tanks from collision or grounding damage, refrigerated LNG ships are constructed with a complete double hull, i.e. an inner and outer hull. The void space between inner and outer hull is utilised for ballast water.

As LNG has a low density of about 0.45 tonnes/m^3 compared with 0.8 to 0.9 for conventional oil fuels, LNG ships have a relatively shallow draft and a large freeboard.

Figure 8.2 **LNG Ship Containment Systems: Main Design Features**

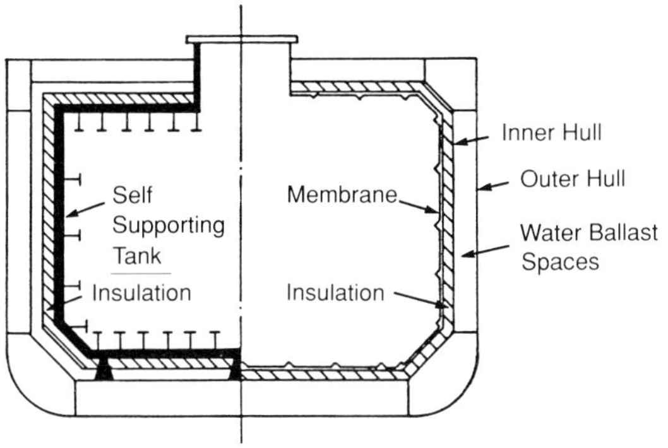

Self Supporting Tanks

Tank – Heavy Rigid Metallic Tank
    High Material & Fabrication Cost

125,000m^3 Ship Tank Material Weight 4,000 tons

Insulation – Relatively Cheap Non-Load Bearing

Membrane System

Tank – Heavy Rigid Metallic Tank
    High Material & Fabrication Cost

125,000m^3 Ship Tank Material Weight 400 tons

Insulation – Rigid Load Bearing Over Whole Surface.
    Relatively Expensive

Types of Cargo Containment Systems

The combination of metallic container and insulation is termed the cargo containment system. Over the years development of containment systems for LNG refrigerated ships has been along two broadly independent but parallel lines, i.e. self-supporting tank systems and membrane systems. The main design features of these two systems are illustrated in Figure 8.2.

Self-supporting tanks are, as their name implies, heavy thick-walled structurally independent metallic cargo tanks. These tanks, apart from being supported along their base, are sufficiently rigid to withstand the hydrostatic cargo forces without assistance from the ship's hull. Because they are structurally strong and require considerable quantities of special cryogenic material, they are high in material and fabrication cost. On the other hand, as the insulation is not required to carry cargo loads, it may be made of relatively low cost materials.

Membrane tanks rely on a light, thin metallic membrane that because of its lightness and flexibility, requires a rigid load bearing insulation system over the entire tank surface that will allow all the hydrostatic cargo loads to be transmitted through the insulation to the hull of the ship. The insulation system must provide rigid support over the whole membrane surface while allowing relative expansion and contraction of the metallic membrane where required. Consequently, although the metallic material weight and cost is significantly reduced compared with the self supporting tank system, this is at least partially offset by the more sophisticated and expensive insulation requirements.

Although the above illustrates the fundamental differences between the two principal design philosophies, detailed design of the components varies significantly from one commercial design to another. Containment systems designed for marine transportation bear little resemblance to shore storage tanks for containment of the same product, although some components and concepts of the system may be common to both. This is because the environment of ships' tanks is very different from that of tanks on land. Ships' tanks, of necessity, are tailored to make the best possible use of existing hull forms in terms of utilisation of hull volume and, in varying degrees, the strength of the hull. Furthermore, shore tanks are static whereas ships' tanks are subjected to additional dynamic effects caused by ship movement in a seaway and external stresses induced into the cargo containment system by bending, and torsional flexing of the ship's hull due to wave forces.

In addition to the two principal design concepts outlined above, other designs exist, at least on paper, such as semi-membrane systems that lie somewhere between the membrane and self-supporting tank systems. However, none of these alternative designs have yet gained commercial acceptance or operating experience.

Metallic Container Materials

As already mentioned, the metallic container, whether it be a membrane or a self-supporting tank, must be constructed from a material capable of maintaining its strength and ductility i.e. flexibility at low temperatures.

For temperatures at around -160°C the designer is forced to use materials which can be say 3 to 15 times more expensive than normal mild steel – see Table 8.1 – subject to variations in world metal prices.

Material cost per unit weight, however, is not enough in itself to act as a criterion for material selection. A better criterion is to relate cost per unit weight, density and maximum allowable design stress into a design cost factor as shown in Table 8.1. On this basis 9 per cent nickel steel and aluminium alloy have relatively low design cost factors. This indicates that designs requiring large quantities of cryogenic tank material, i.e. self-supporting systems, should use either of these two materials. In practice, all self-supporting tank systems built to date use either one or the other, although new materials are under development that in future may give the same performance at lower cost. Similarly, stainless steels and Invar have special application for membrane designs where, although the material cost is high, the material itself and the method of construction allow it to be used in very thin sections.

Table 8.1 **Cryogenic Material and Design Cost Factors**

Material	Relative Material Cost	Design Cost Factors*	
Mild steel (as a base)	100	1.0	
9% nickel steel	250	1.6	Self supporting tanks
5083 aluminium alloy	250	1.6	
304 stainless steel	500	4.5	Membrane tanks
36% nickel steel (Invar)	1250	7.5	

$$* \text{ Design cost factor} = \frac{\text{material cost} \times \text{density}}{\text{design stress}}$$

Insulation Materials

Materials for cargo tank insulation should generally have the following properties:

- *low thermal conductivity*
- *resistance to the cargo product*
- *low flammability*
- *low material and installation cost*
- *suitability for the lowest temperature anticipated*
- *capability of withstanding thermal stresses in service*
- *resistance to moisture and sea water*
- *high chemical and mechanical stability*

In addition, for membrane designs and for the bottom tank support for self-supporting tank designs, they must be capable of withstanding the loads due to the weight of tank and contents and any dynamic hydrostatic cargo loads involved.

The materials commonly used for these duties are:

Non-load bearing – Fibreglass   Load bearing – High density
                  Perlite   foam materials
                  Low density   Balsa wood
                    foam materials   Plywood

Secondary Barriers

The metallic container, which is in direct contact with the cargo, whether it is a metallic membrane or self-supporting tank, is generally designated as being the primary barrier or primary container. For LNG, a secondary barrier may be required to act as a temporary container for any leakage of liquid cargo through the primary barrier.

The requirement for, and the extent of, the secondary barrier depends on the type of primary barrier. In general, all membrane designs require a secondary barrier over the whole surface area of the tank. Self-supporting tanks on the other hand, have secondary barrier requirements varying from complete to no secondary barrier at all. Often, rectangular self-supporting tanks require a partial secondary barrier extending over the lower regions of the tanks whereas spherical tanks, which can be subjected to detailed stress analysis relatively easily, do not require any secondary barrier. For designs requiring no secondary barrier, the inner bottom hull steel under the cargo tank is required to be protected against minor liquid cargo leakage by some form of 'drip tray'.

Ship Construction Regulations

Various designs have evolved over the years, some of which are described below. Early designs were based on construction codes of practice issued by various national authorities. These were intended to provide standards of construction to minimise the risk of failure with consequential adverse effects on the ship, its crew and the environment. While the aims of these codes were identical, the means of achieving these aims in terms of design criteria varied in detail from one code to the next. To rectify this situation for liquefied gas carriers, the International Maritime Organisation (IMO) published in 1976 a code to provide an international standard for the safe carriage by sea of liquefied gases by presenting the design and constructional features of ships involved in this trade and the equipment they should carry. The code gives requirements on location of cargo tanks, design criteria for cargo tanks, allowable materials, insulation systems and environmental control. The IMO code, which is reviewed and updated as necessary, is now the "bible" of liquefied gas carrier design and its requirements are enforced by the governmental bodies concerned in every major maritime nation of the world.

Self-supporting Tank LNG Designs

IHI-SPB

This Japanese system, while appearing under this name in recent years, is essentially an updated and improved version of the CONCH rigid tank design that was the first design to enter commercial service in the early 1960s with the Algeria-UK project (METHANE PROGRESS and METHANE PRINCESS).

The IHI-SPB design (Figure 8.3) utilises prismatic aluminium tanks that may be rectangular or trapezoidal as best suited to their location in the ship. The flat walls of the tank are internally stiffened by webs and girders to reduce the stresses and deflections to acceptable limits under service conditions. The tanks are fitted internally with a longitudinal, liquid-tight bulkhead and a transverse swash bulkhead to reduce the effects of cargo free surface and liquid surge motions respectively.

Figure 8.3 **IHI-SPB Free-standing Tank System**

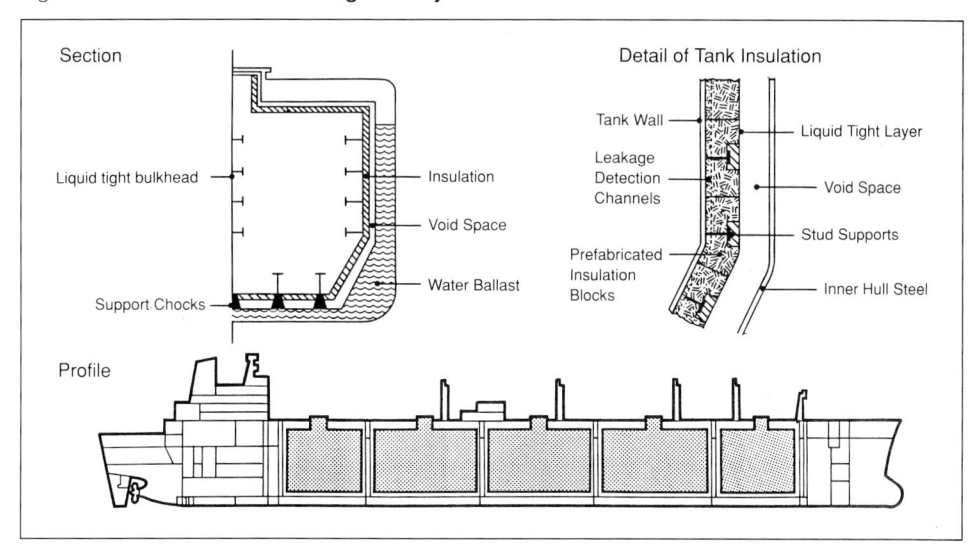

The tanks are supported in the ship's hold on several rows of load bearing insulating pads, which also serve to locate the tank under ship rolling conditions. The insulation system installed on the external surface of the tanks consists of prefabricated composite foam panels held by studs welded to the tank walls. Preformed channels in the insulation adjacent to the tank allow these spaces to be monitored for cargo vapour leakage and also guide any liquid leakage to drip trays under the tank. There is access space around all sides of the tank insulation to allow inspection during construction and subsequent maintenance.

Only one ship of this (CONCH) design was in service in 1991. However, two IHI-SPB design ships were under construction.

Kvaerner-Moss

While the first two LNG ships to this design (Figure 8.4) utilised spherical cargo tanks constructed of 9 per cent nickel steel, all subsequent ships have used aluminium alloy. The sphere contains no internal structural members or bulkheads. The support is a metallic cylindrical skirt of a suitable material welded to a specially shaped section that forms the equator of the sphere. The skirt must have adequate strength to withstand the hydrostatic and dynamic loads at the minimum anticipated operating temperature. The bottom of the skirt is welded to the ship's structure in a manner suitable for both the inner hull and skirt materials. The skirt deflects when the tank contracts and limits stress in the tank shell. A main requirement of this design is that it satisfies the "leak before failure" concept which assumes, because of the high level of stress analysis possible with this type of tank, coupled with extensive non-destructive testing during fabrication, that if a crack does occur it will grow very

Figure 8.4 **Kvaerner-Moss Free-standing Spherical Tank System**

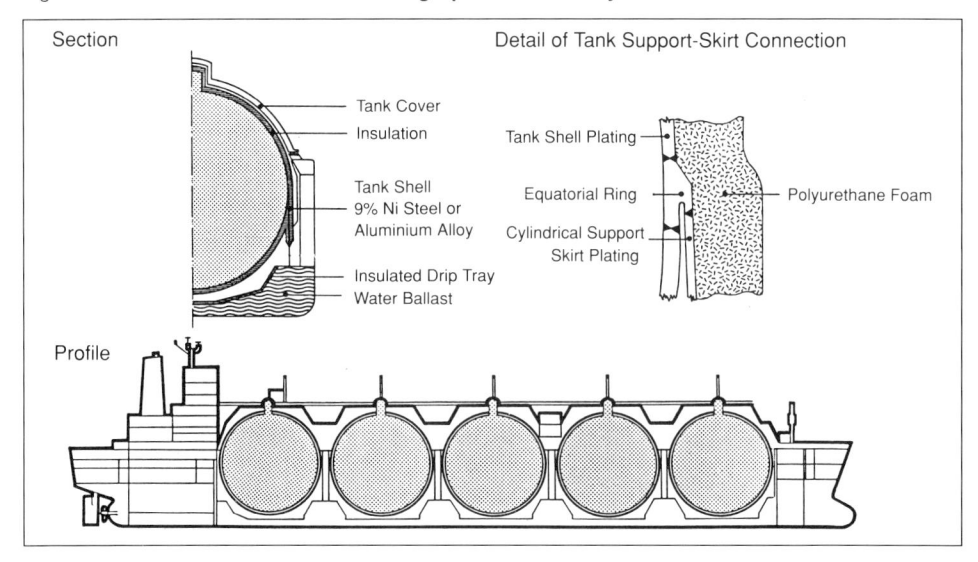

slowly and be detected as a small leak before the crack reaches serious proportions. Meeting this requirement permits the secondary barrier to be reduced to the extent of a drip tray under the tank.

One form of insulation currently employed consists of polystyrene blocks (3 m long, 300 mm thick) thermally welded together to form a continuous spiral string of insulation applied to the outer surface of the sphere. The polystyrene foam is then covered by an

Figure 8.5 **LNG Ship Construction, Nagasaki, Japan**

With the bottom and mid-section of a 40m diameter sphere in place, the top section is about to be lowered into position. Courtesy of Woodside Offshore Petroleum Pty. Ltd.

Figure 8.6 **'Northwest Sanderling' Berthing at Sodegaura, Japan**

This ship of 125000 m³ capacity built by Mitsubishi Heavy Industries to the Kvaerner-Moss design entered service with the Australia North West Shelf project in 1989.

aluminium outer cover for fire protection and to act as a vapour barrier. Insulation is also applied to both sides of the cylindrical skirt. A space exists between the insulation and the tank to allow circulation of inert gas or dry air with a series of channels that will allow any LNG leakage to drain to the drip tray. An alternative insulation system of prefabricated composite foam panels held by studs welded to the sphere is also used.

Of the 65 ships in service in 1991, 31 are of this design concept.

Membrane Tank LNG Designs
Technigaz
In this design (Figure 8.7) the membrane is made of 1.2 mm thick stainless steel having two series of corrugations perpendicular to one another. These corrugations absorb the thermal contraction and ship's strains. The membrane is welded to corner angles which form the end restraints and is attached in a regular pattern to the insulation panels. The original insulation system consisted of wooden grounds bolted to the inner hull to provide a support for the insulation panels. The spaces between grounds are filled with fibreglass. The insulation panels themselves are made up from prefabricated balsa wood sections glued together with the joints between adjacent panels sealed by PVC wedges and plywood seal splices. Both faces of the balsa wood panels are covered with plywood sheets, the inner (cold) plywood face providing the system's secondary barrier. Further balsa pads are attached by

glueing them to the inner plywood face to provide a flat surface upon which the membrane is fastened and supported. At all hold corners, hardwood keys are provided to anchor the membrane through heavy stainless steel angles.

The membrane corrugations and the fibreglass spaces are kept under an inert atmosphere in service.

Figure 8.7 **Technigaz Membrane Tank System**

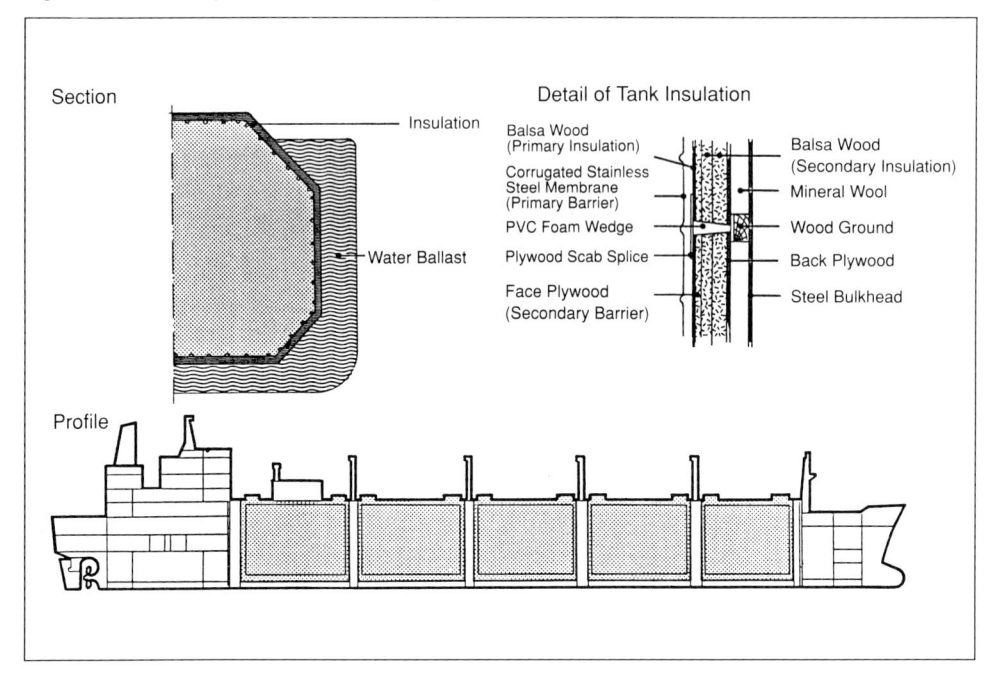

While this system is representative of ships to this design actually in service, later refinements have replaced the balsa wood panels by PVC/PUF foam sections in an effort to reduce the costs and complexity of the system.

In 1991, there were 8 ships of this design in service.

Gaz Transport
This system consists of virtually two identical membranes and insulation systems, one within the other, to form a composite two barrier system (see Figure 8.9). Current systems use a 0.7 mm thick Invar membrane for both primary and secondary barriers. Because Invar (a 36 per cent nickel alloy) has a very low thermal expansion coefficient, the thermal contraction in service is small and there is no requirement for corrugations, etc.

The primary and secondary insulation consists of a multitude of plywood boxes filled with perlite powder insulation, similar to large building bricks, attached by studs and bolts through special framework to the inner hull initially. The secondary

Figure 8.8 **Interior View of a Technigaz Cargo Tank**

BEBATIK, total capacity 75000 m³, built by Chantiers de l'Atlantique, France. An impression of the size of the tank is shown by the man standing at the forward end.

membrane is held in place by special Invar "tongues" welded to the membrane and fixed to the plywood box faces. The primary insulation boxes are fixed in place by a series of vertical channel sections of 9 per cent nickel steel anchored to the wooden framework through the secondary barrier. Again, Invar "tongues" attached to the plywood box faces form a welded attachment and support for the primary Invar barrier.

In common with most membrane systems, special means are normally provided to support cargo pumps, piping, instrumentation, etc., other than by direct attachment to the membrane. This is generally in the form of a tower framework suspended from the deck trunk and restrained from horizontal movement at the tank bottom.

Twenty ships built to this design were in service in 1991.

Internal Insulation Systems

It has been the designer's dream from the early days of LNG transportation to develop an insulation system that could function not only to limit heat transfer to the cargo and protect the hull steel, but which would also be capable of acting as the primary cargo container direct, i.e. without requiring any additional metallic barrier. Over the years extensive research has been applied in an attempt to develop such a system but none have yet reached the stage of commercial viability.

LNG Ship Designs

Of the many LNG ship designs that have been proposed, only the four containment systems outlined above can be considered as being technically and commercially

Figure 8.9 **Gaz Transport Membrane Tank System**

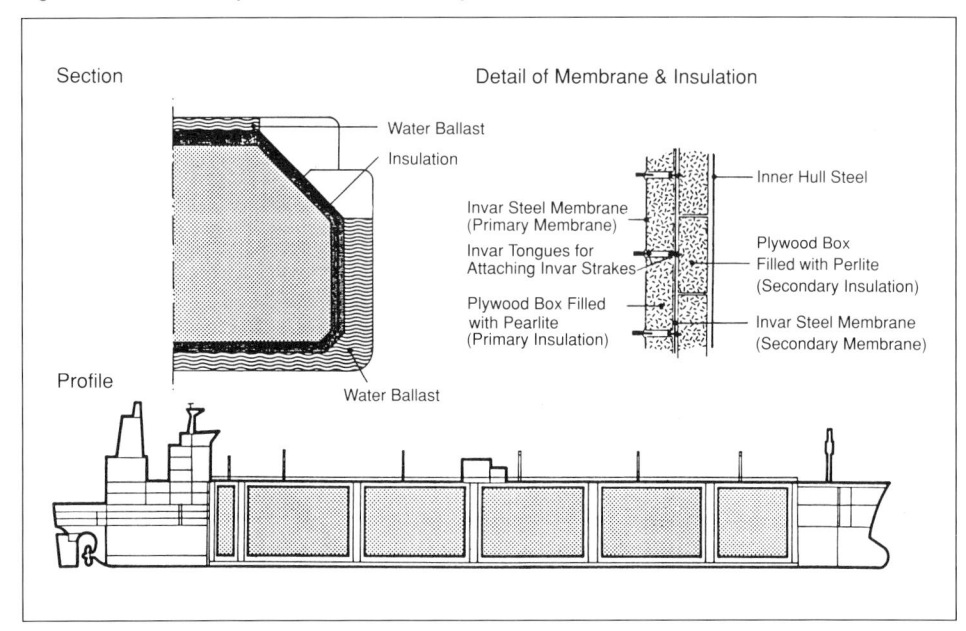

Figure 8.10 **Two Gaz Transport Membrane Type Ships**

LNG BONNY and LNG FINIMA, capacities 133000 m^3 each, built in 1981 and 1985 respectively by Kockums, Sweden, being refurbished after lay-up by Mitsubishi Heavy Industries in Yokohama, Japan.

approved, principally because only these systems have proved their performance and reliability over long periods of operation. In the case of the IHI-SPB prismatic system, this includes experience with the almost identical CONCH system.

These established designs, although continuously under development in terms of detail improvements, have largely realised their potential. No major changes or improvements can be expected in the future and as such they are now the accepted standards against which any other design must be judged. Also in many cases shipyards specialising in LNG ship construction have adopted one or more of these designs as being particularly suitable for their own facilities. Moreover, they have invested considerable capital to provide the necessary special equipment and expertise required for the system concerned. Consequently, a new system is only likely to achieve commercial success if it can convince shipyards and shipowners that:

- *it has significantly lower capital and/or operating cost*
- *it simplifies construction and shipyard facility requirements*
- *it has exceptional operating advantages*
- *it has significantly higher safety features*

Given the very high cost of development in terms of research effort, manpower and time required to engineer and obtain regulatory approval for any new design, the likelihood of a radical new design seriously challenging these established designs is remote.

Finally, it is not appropriate to suggest which might be the preferred design, each have their advantages and drawbacks compared with the others and prospective owners will need to decide themselves which design best meets their particular requirements. However, all the systems described should be considered equally safe, complying with IMO, regulatory and classification rules.

Table 8.2 **Percentage Breakdown of LNG Ship Costs**

Components	Rigid Tanks	Membrane Tanks
Hull, outfit & machinery	60	64
Cargo tanks	25	15
Cargo tanks insulation	5	10
Cargo handling system	10	11
	100	100

Cost comparisons are not meaningful as these are constantly changing reflecting, inter alia, the general ship-building industry's situation and that of individual shipyards in particular at any point in time. However, as a very rough guide experience over several years suggests that rigid tank systems tend to be marginally more expensive than membrane systems; Table 8.2 gives an approximate cost breakdown of the two systems.

Ship Cargo Operations
Regardless of the type of containment system employed, heat transfer will take place from the ambient air and sea water into the cargo through the insulation. As LNG is carried at or very near its boiling point at atmospheric pressure, any heat added to the cargo will cause part of the cargo to evaporate. This vaporised cargo is termed "boil off". The quantity of boil off generated will depend on the thickness and effectiveness of the insulation system, the surface area of the cargo tanks, ambient temperature conditions, the movement of the ship and numerous other factors.

In practice, all containment systems are or can be designed for average boil off in the loaded condition in the range of 0.1 to 0.3 per cent of cargo capacity per day. The larger the ship, the lower the boil off in percentage terms. The upper end of the range applies mainly to smaller ships built in the 1970s with early insulation systems, whereas the lower end refers to the minimum achievable with larger new ships fitted with improved insulation. The design boil off may be greater than the minimum achievable for reasons of economy of insulation and to allow the creation of sufficient boil off to use as propulsion fuel.

On the ballast voyage, when there is generally only a little cargo remaining in the tank bottoms, the boil off is about 30-70 per cent of the loaded voyage quantity, depending on operational requirements and practices for the particular containment system.

Using the boil off as fuel is the standard practice on all LNG ships today. This is mainly because reliquefaction plants are complex, expensive in both capital and operating cost, and hence have not been shown to be economic. Careful design, voyage planning and operation using the boil off as fuel can reduce the ship's propulsion fuel oil requirement to virtually zero on the loaded voyage and to 40-50 per cent on the ballast voyage. In most cases the boil off is burnt as fuel in the main boilers of a steam turbine propulsion plant, although it could also be used in dual-fuel diesel engines or gas turbines.

If boil off is burnt as fuel in excess of that required for propulsion of the vessel and other steam services, then the excess steam can be "dumped", i.e. condensed using a seawater cooled condenser. The amount of excess gas burnt can be controlled to match the boil off available. As no benefit is gained from the steam created in such a manner, it is considered as a direct product loss. This system is now frequently fitted in addition to venting capability, but used in preference to venting, as restrictions on venting may be imposed by port administrations, whereas "dumping" can take place at any time.

Cargo Transfer Operations

During loading any excess boil off generated on the ship is returned to the shore by compressors, installed either on the ship or shore, where it may be used as fuel in the shore power plant or reliquefied and returned to the shore storage. The vapour return flow rate will normally be highest during the initial stages of loading when the pipelines and receiving tanks are relatively warm. During this phase the loading rate will be restricted so that the vapour return flow rate is limited to the available capacity of the compressors. As the loading system temperature stabilises the vapour flow rate will reduce and the LNG loading rate increase. Loading systems using dedicated ships are usually designed to allow a loading operation of a large LNG ship to be completed within about 14 hours.

Discharging is essentially the reverse of loading. Excess boil off generated in the receiving shore tank by the ship's cargo pumps and pipeline system will be partially returned to the ship to replace the liquid volume displaced. The remainder can either be used within the terminal and/or fed into the terminal's send out line. When discharging LNG it is normal to retain some cargo for the ballast voyage. This ensures that the cargo tanks are kept reasonably cold on the ballast voyage so that loading can commence at the loading terminal without wasting time in cooling down the system. The quantities retained will depend on the length of voyage, system cooling requirements, the value of the product, etc.

Table 8.3 **Typical Voyage Losses Percentages of Ship's Loaded Cargo Capacity**

Losses	Low	High
During loading	0.5	1.5
Boil off during loaded voyage (10 days)	1.0	3.0
Boil off during ballast voyage	0.3	1.5
Total	1.8	6.0
Ratio of LNG delivered : ex-storage	98.2	94.0

Safety

Safety is of paramount importance in the design, construction, operation, maintenance and repair of LNG ships. Some of the more important design features, e.g. the insulation systems and double hulls to reduce the likelihood of cargo spillage in the event of collision or grounding of the ship, have been discussed. These are complemented by inbuilt systems for the early detection of any cargo leakages, to limit the quantity of any spilled cargo and facilitate its safe disposal. Careful material selection and thorough stress analysis is carried out at all design stages with rigorous quality control being exercised throughout construction. Extensive fire protection and prevention systems are provided.

The selection and training of officers and crew for normal operations, and to handle any emergencies that may arise, receives high priority, as it will for those personnel (tug crews, harbour staff and shore-based operators) concerned with the entry, berthing and departure of LNG ships at the loading and receiving terminals.

Detailed procedures for the purging of the ship's tanks, pipes and related cargo equipment are laid down to ensure that any maintenance or repair work is carried out in a safe, gas-free environment.

All these aspects of design, construction, operation, etc., backed by an extensive on-going research programme and practical trials on the behaviour of LNG spills under a variety of conditions, have contributed to the excellent safety record the LNG shipping industry has enjoyed over the last 30 years. Experience to date indicates that provided an LNG ship is operated and maintained to the highest possible standards, it can have a useful, safe service life of some 40 years, perhaps longer.

Ship Design and Fleet Configuration Considerations

The majority of LNG ships in current service were designed and purpose-built for long term continuous service in a specific project. A number of ships were also built mainly in the 1970s on a speculative basis by independent owners; some of these ships have since been scrapped or converted. Some others, after long periods of lay-up, have been refurbished and added to existing project fleets to provide extra capacity on a short or long term basis for projects which, one way or another, have increased their liquefaction output above their original contractual commitments. Although such ships may not necessarily be the optimum size, speed, etc., this is usually more than compensated by their lower capital cost, including the cost of refurbishment, compared with the current cost of a newbuilding which is now well in excess of US$200 million for a 125000 m^3 capacity ship.

For a new fleet for a new project, assessing the number, size, speed, etc., of the ships to be provided will include the following considerations:

- *the quantity and regularity of the annual contracted volumes to be delivered;*
- *the voyage lengths involved and hence the amount of cargo 'lost' through boil off;*
- *the time to be set aside for regular dry docking, maintenance and surveys by, for instance, classification societies and government authorities;*
- *scheduled delays, e.g. slow steaming through congested waterways;*
- *unscheduled delays due to breakdowns, bad weather, etc., based on experience from ship operations in that geographic area;*
- *port times for harbour steaming, queuing, berthing, loading and unloading, possible restrictions on ship movement at low tide or in darkness; and,*
- *draught limitations and multi-port discharge, if appropriate.*

Computer programs are available for resolving these and similar considerations to assist the project analyst in arriving at the optimum specification and number of ships for the trade in question. However, such assessments cannot be made in isolation and full account has to be taken of any supply constraints, e.g. liquefaction plant downtimes for maintenance, etc., and buyer's requirements such as having to dovetail with deliveries of LNG from other sources.

Ship Ownership and Charter Parties

A project fleet, or parts thereof, may be owned directly by the LNG seller, by a separate entity established by the seller for that purpose, by an independent shipowner, or by the LNG buyer. The preferred route for any particular project will be conditioned by a variety of factors including whether the sale is FOB, CIF or ex-ship, and the risks, benefits and reponsibilities involved of a corporate, contractual, financial, fiscal, economic, technical and operational nature.

If, as in most instances, the ships are owned by a separate corporate entity or entities, even if the latter are in turn wholly owned by the LNG seller or buyer, the ships will be 'locked' into the project by a separate contract, i.e. a charter party, which can take several forms.

In the case of BAREBOAT or DEMISE CHARTERS, the owner carries no 'off hire' risk and is responsible for the design, construction and capital costs of the ship, while the charterer hires the ship for a specified period, manages the ship and pays for all crewing, dry docking, repairs, insurance and other operating costs. The charterer thus has total control over operating the ship and at the end of the charter must return it to the owner in the same condition as on acceptance, fair wear and tear excepted.

A TIME CHARTER, as the name implies, covers the hiring of a specific ship for a specified period, but unlike a bareboat charter the owner must maintain the ship in satisfactory condition throughout the charter and pay the cost of crewing and running the ship including insurance. The charterer is responsible for bunkers, port charges and dues, etc., but does not pay the hire charge when the ship is out of service for docking, breakdowns and strikes by the owner's personnel. However, the charterer has to pay for any replacement ship he may decide to hire.

CONSECUTIVE VOYAGE CHARTERS are where the owner and charterer agree terms for the use of a specific ship for a specified time or alternatively for a specific number of voyages. Full responsibility for ship operation lies with the owner, the charterer only being responsible for issuing loading and discharge instructions. A certain time (laytime) is allowed for in each port and if this is exceeded the owner can levy a penalty (demurrage). The owner does not have to provide a replacement in the event of breakdown. Voyage or Spot Charters are very similar but of shorter duration or for single voyages only.

CONTRACTS OF AFFREIGHTMENT are where the shipper undertakes to provide a given volume of LNG over an agreed period and the owner undertakes to nominate ships to lift cargoes (within an agreed range) at certain intervals throughout the contract period. Laytimes and demurrage rates are specified and nominating procedures agreed. In the event of breakdown, the owner may have to provide a replacement except after loading when the shipper may be required to arrange that himself. Although the shipper is completely free of the problems of ship operating, he pays the owner a penalty if he fails to provide sufficient cargoes.

The foregoing descriptions are not comprehensive but simply indicative of some of the principal differences between various forms of charter parties.

Chapter 9

LNG RECEIVING TERMINALS

The main purposes of an LNG receiving or regasification terminal are to receive LNG deliveries from ocean-going LNG carriers, to store the LNG, to regasify the LNG and to send out gaseous natural gas as and when required. Some terminals also have facilities for loading road or rail trucks which deliver LNG to smaller 'satellite' storage and regasification stations located elsewhere in the marketing area.

Whereas LNG receiving terminals are sited on or close to the coast, or occasionally alongside a waterway connected to the sea, satellite stations are almost invariably located inland. Depending on local market circumstances, an LNG terminal may have a base-load role sending out gas throughout the year, although actual daily send-out rates can vary significantly. Alternatively, it may have a seasonal or peak-shaving role and only send out gas for a few months, weeks or even days of the year. The majority of terminals currently in service are fundamentally base-load terminals.

Main Components
An LNG terminal consists of the following components:

- *LNG carrier berthing and unloading facilities*
- *LNG storage*
- *a regasifying or vaporising system*
- *facilities to handle boil off gas*
- *high pressure LNG pumps*
- *metering and pressure regulation station*
- *gas delivery pipeline*

In addition, gas odorization, calorific value control and LNG road/rail loading facilities may be provided. Some terminals, mainly in Japan, are also equipped with LPG unloading, storage and vaporisation facilities so that LPG can be blended in to raise the calorific value of the vaporised LNG.

Layouts of terminals vary as they will be conditioned by a variety of local factors including the plot area available, its proximity to deep water and residential areas, environmental considerations, the necessity to conform with safety regulations, the type and amount of storage required, etc. A photograph of a terminal is shown in Figure 9.1 and a typical simplified flow scheme in Figure 9.2

Site Selection Criteria
The selection of a suitable site for an LNG receiving terminal requires the following conditions to be investigated and satisfied:

Figure 9.1 **Tokyo Gas's LNG Terminal: Sodegaura**

In the foreground are 3 above-ground and 17 in-ground tanks. Behind in Tokyo Electric's Power Station with its storage to the right. Courtesy of Diamond Gas Operation Co. Ltd.

Figure 9.2 **LNG Terminal : Simplified Flow Scheme**

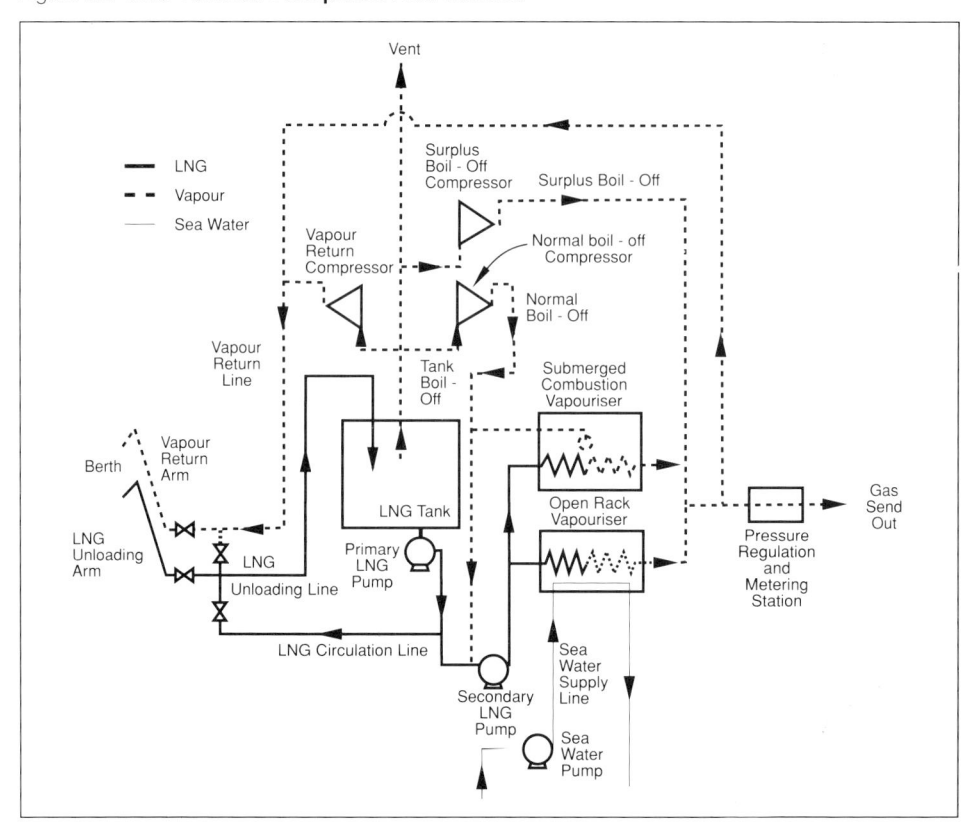

LNG carrier approach	– *water depth, width, alignment, currents, navigational aids, other marine traffic, tugs and pilots.*
Manoeuvring basin	– *water depth, space, wind, waves, currents, tugs, pilots and other marine traffic.*
Berth	– *water depth, wind, waves, currents, sterile zone, jetty length, mooring lines, tugs, pilot and other marine traffic.*
Site	– *land availability, site preparation, onshore access, utilities and emergency services.*
Pipelines	– *their route, nature (onshore and/or offshore), geology, topography, depth and length.*
Environment	– *protected areas and current usage of the area.*
Safety	– *separation from other marine traffic, industry and populated areas.*
Cost	– *the total cost of the terminal to the point where gas is delivered ex-terminal.*

Some of these considerations are self-explanatory, others are discussed in more detail later.

Berthing and Unloading Facilities

A berth, or berths, will either be provided alongside the terminal, if sufficient water depth is available (or can be dredged) to accept the LNG carriers that will service the terminal, or if not, a trestle or jetty may need to be built to connect the berth to the shore. The length and cost of the trestle is balanced against the cost of dredging to reduce the trestle length. In one instance, Cove Point, Maryland, the berth is connected to shore by a pipeline laid in an under-sea tunnel.

Each berth is equipped with several unloading arms which connect the ship's manifolds with the onshore manifolds. These incorporate swivel joints to allow for ship movements. Unloading lines from the arms run back to storage. These lines are usually made of stainless steel or some other material, e.g. aluminium, which can safely withstand the extremely low temperatures of LNG. They will be insulated to reduce as far as is practicable heat transfer from the atmosphere to the LNG.

Most terminals have more than one unloading line so that the lines can be kept cold between cargo deliveries by continuously circulating LNG from storage to the berth and back again. If only one unloading line is installed, then a separate circulation line has to be provided to feed LNG into the unloading line to keep it cold.

Separate vapour return lines have to be provided to carry vapour from the terminal back to the carrier to fill the space vacated by the LNG as it is unloaded. Alternatively, in the absence of a vapour return line, the ship's vaporisers have to vaporise some of the cargo for this purpose.

Figure 9.3 **LNG Unloading Arms: Tokyo Gas's Sodegaura Terminal**

Courtesy of Tokyo Gas Co. Ltd.

Types of Storage

As the same concepts apply to storage for liquefaction and peak-shaving plants, these are discussed collectively in Chapter 7.

Pumps

Because the capital and operating costs of LNG pumps are lower than those of gas compressors, LNG is pumped at high pressure, typically 50 to 80 bar, into the vaporisers so that the resulting gas needs no further compression.

Low pressure LNG circulation pumps are installed inside or close to the LNG storage tanks. LNG delivered by pumps is circulated through the unloading line(s) as described above and to the high pressure pumps which raise the pressure to a little above the gas send-out pressure.

Vaporisers

The function of the vaporisers is to warm up the LNG so that the resultant gas is at or above 5°C. There are several types of vaporisers in all of which LNG flows through tubes or panels which are heated in various ways from the outside.

So-called open rack vaporisers consist of panels which draw heat from sea water to warm the panels. Large panels and large quantities of sea water are required because water can only be cooled through a few degrees to avoid excessive build-up of ice on the panels. Ice acts as an insulator and reduces the rate at which LNG can be vaporised.

Sea water may need to be warmed in winter. Equally, the discharge of cold sea water back into the sea can give rise to environmental concerns on the possible effect this may have on marine life. To overcome such problems, some vaporisers employ an 'intermediate fluid' to warm the LNG in the vaporisers. This fluid then flows to a second heat exchanger where it is warmed by sea water before being returned to the vaporisers. Efficiencies can sometimes be improved if the LNG terminal is able to draw on the warm cooling water discharged from a nearby power station.

Some open rack vaporisers work on the principle of drawing heat from the surrounding air instead of from sea water.

Figure 9.4 **Open Rack Sea Water Vaporiser.**

Courtesy of Diamond Gas Operation Co. Ltd.

The second main type of vaporiser operates above ambient temperature and is known as a heated or submerged combustion vaporiser. LNG is passed through a tube bundle immersed in a water bath which is heated by gas, electricity, LPG, hot water or steam. Although these vaporisers are smaller and cheaper than the open rack type, they are more expensive to operate because of the energy (fuel) they consume. Typically, a gas-fired vaporiser can consume between 1.5 and 2.5 per cent of the throughput.

Most terminals are equipped with both open rack and submerged combustion vaporisers, the combination and number of vaporisers provided reflecting the main role of the terminal in relation to the market in question, the quality and temperature of sea water available, ambient temperatures during the year and similar factors.

Some terminals use a suitable fluid to warm the vaporiser which, having been chilled by the LNG in the vaporiser, can be used for cooling purposes in a nearby air separation plant, food freezing factory, chemicals plant, etc. The warm fluid is then returned to the terminal. Fluids used in this cold loop can range from brine or Freon for temperatures above -40°C, to nitrogen where very low temperatures are required.

Boil Off Gas Facilities

As no insulation system is perfect, heat leaks slowly and continuously through the insulation of LNG tanks, lines and other equipment with the result that some LNG boils off as vapour. The total daily rate of boil off for the whole terminal is normally less than one per cent. This boil off can be compressed in a reciprocating or centrifugal compressor and mixed with the LNG stream in an absorber installed between the primary and secondary LNG pumps and/or fed into the output gas stream from the vaporisers. Some of this boil off may be used as a fuel if the terminal is equipped with gas-fired vaporisers and/or to generate electricity for power uses within the terminal.

When LNG is being discharged from the ship into the terminal's tanks, it displaces vapour from the tanks. Some of this vapour can be returned via a compressor to the ship's tanks. However, as vapour is also boiling off in the ship's tanks, the volume of vapour returned to the ship must be less than the volume of LNG delivered into the terminal's tanks.

This 'surplus' boil off vapour, and that created as LNG is being unloaded and warmed up by work done in the ship's pumps, also vapour resulting from heat leakage into the unloading arms and lines, can be more than the terminal requires for various operational purposes. In this event, the surplus boil off is compressed and fed into the terminal's outlet system downstream of the vaporisers.

If it is possible to match the designs of the terminal and ships so that the pressure in the terminal's tanks is 70 to 100 mbar (1 to 1.5 psi) above that of the ship's tanks, then surplus boil off is reduced considerably.

To avoid malfunctioning, separators are normally installed at inlets to LNG pumps, and elsewhere as may be necessary, to ensure that any vapour that may have been formed in unloading lines, etc., cannot enter the pumps. Likewise, knock-out drums may need to be fitted to vapour lines to ensure that any liquid which condenses cannot enter the vapour compressors.

For some terminals used mainly for seasonal or peak-shaving purposes, it may be economic to install facilities to reliquefy boil off during periods when no gas is sent out.

Finally, all terminals are provided with a vent or flare stack to dispose of boil off should there be an equipment failure or if the rate of boil off exceeds the capacity of the terminal's compressors.

Metering and Pressure Regulation Station

Gas before leaving the terminal passes through a pressure regulating and metering station and in most instances will be odorised.

Safety

Safety considerations are always of paramount importance, especially so for LNG terminals which are frequently located in or close to major conurbations. The design of the terminal and the equipment and materials used will be focused on preventing any spills of LNG or leakages of gas, and in the unlikely event of this happening, ensuring that they are dispersed safely.

For above ground tanks, retaining walls will be erected around each tank of sufficient size to contain its total content in the event of the tank failing. These walls must also be capable of withstanding low temperatures. For obvious reasons retaining walls are not necessary for tanks which are totally in-ground but both types of tank will require to be surrounded by gas and fire detectors and fire fighting equipment.

Distances between and the locations of tanks, vaporisers, liquid and vapour lines and other facilities, will be determined by local regulations and/or by recognised international codes of practice.

Control rooms, workshops, offices and other buildings will be sited in safe locations. Means to ensure that unauthorised people cannot access operational areas of the site have to be established, while the arrival, departure and berthing of LNG carriers will normally be controlled by the local harbour authority or by the operating company as appropriate. These and other safety precautions will tend to be site specific including the need to satisfy the requirements and regulations of the local authorities concerned.

Gas Quality

Quality can be an important consideration if the terminal is supplied with LNG from more than one source and/or if the marketing area it serves receives supplies from other sources, indigenous or imported. In order to ensure that the calorific value and Wobbe Number of the gas supplied to end-consumers are constant and compatible within prescribed limits, it may be necessary to dilute or enrich the vaporised LNG before it leaves the terminal.

In Japan, which imports LNG from a variety of sources, it is common practice for gas utilities to enrich imported LNG with LPG to raise the overall calorific value to a common level higher than that of vaporised LNG. Variations in gas quality are not so important, however, where all or most of the vaporised LNG is fed directly to a large consumer such as a power plant.

Storage Requirements

The following comments assume that imported LNG is to all intents and purposes the sole source of gas supply for the market in question and that the terminal is therefore required to perform other duties beyond that of just receiving, storing and regasifying LNG. In other instances, where LNG imports are only a supplement to other sources of gas supply, or where seasonal/peak-shaving considerations do not arise, the assessment of storage requirements may be a less complex process.

Almost all terminals are supplied on a regular basis throughout the year, the frequency of which can vary from a cargo every day or two to once every 4 to 6 weeks. Many LNG carriers, but not all those in current service, have cargo capacities between 125,000 and 130,000 m^3 of LNG – see Chapter 8.

Obviously, the terminal must have sufficient storage capacity to accept a full cargo of LNG – the 'replenishment quantity'. If there is a large variation in send-out between summer and winter for markets with a substantial heating demand, then 'seasonal storage' will be needed to cope with a winter of average severity.

Additional capacity may be needed to satisfy demand during the coldest winter which is likely to occur. And there will be an unpumpable stock when the level of LNG falls to the point at which the primary LNG pumps can no longer deliver. It is also necessary to ensure there is sufficient stock to keep the whole system cold. The sum of these is the 'minimum stock'.

The terminal must also be capable of coping with fluctuations in receipts and send-outs. 'Buffer stock' is the volume needed if the send-out should exceed expectations for the ambient temperature in question, or if a delivery is delayed due to bad weather or mechanical breakdown.

'Buffer ullage' is the space needed to allow a cargo to be accepted if the send-out is less than expected or if the ship arrives early.

The total storage requirement is, therefore, the sum of:

- *replenishment quantity*
- *seasonal storage*
- *minimum stock*
- *buffer stock*
- *buffer ullage*

Table 9.1 **Preliminary Calculation of Base Load Terminal Storage Capacity**

Throughput 10⁶ m³ LNG/year	Daily Load Factor %	Replenishment Quantity 000 m³	Buffer Stock + Ullage m³	Seasonal Stock 000 m³	Minimum Stock (Average Year) 000 m³	Total Capacity Needed 000 m³	Number and Size of Tanks Selected x 000 m³
2	50	75	one cargo (0.833 stock + 0.167 ullage)	800	56	1,006	10 x 100
		120				1,096	11 x 100
	85	75	ditto	62	33	245	3 x 90
		120				335	4 x 90
11	75	120	ditto	788	203	1,231	12 or 13 x 100
		200				1,391	14 x 100
	85	120	ditto	554	180	974	10 x 100
		200				1,134	11 or 12 x 100

Notes: 1. The final choice of number and size of tanks should be made by using a simulation model to optimise the liquefaction plant size, number and size of ships and the storage capacity at both the liquefaction plant and the terminal.
2. The seasonal stock level is affected more by the shape of the send-out graph than by the load factor (see Figure 9.6).

This volume, rounded off, will give the number of tanks of a given size which will need to be installed.

The total storage capacity installed in a terminal is usually determined by using a simulation model so that the seasonal, minimum and buffer volumes need not be determined separately. However, an approximate value for each component may need to be calculated initially to determine the range of storage capacities to be used as input parameters to the computer model.

Figure 9.5 **Variation of LNG Stocks in Average Year**

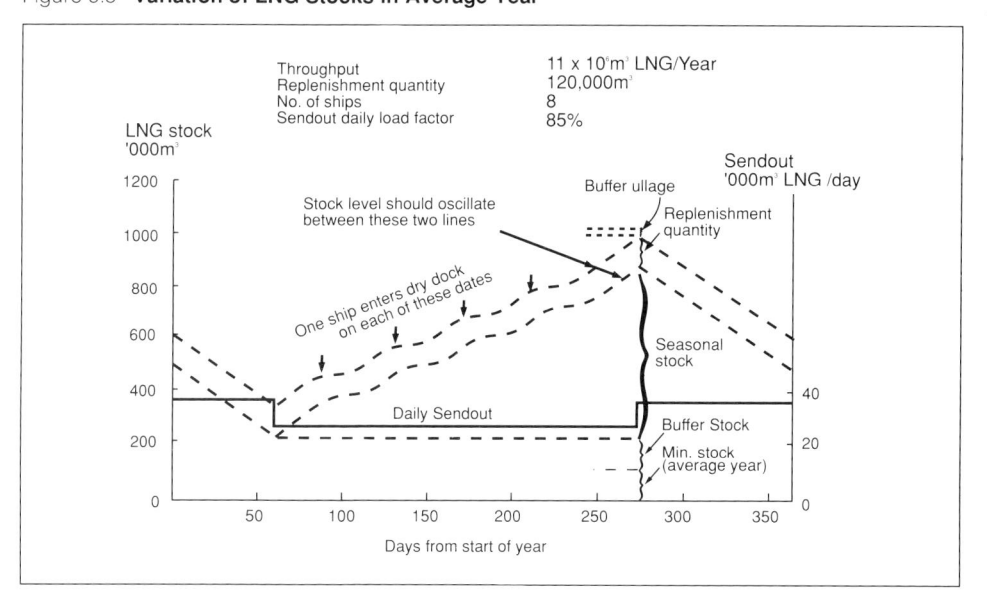

Page 123

Some examples of these preliminary calculations are given in Table 9.1, while Figure 9.5 shows the variation in LNG stocks during an average year for one of these terminals. The resulting stock level is shown on Figure 9.5 as two parallel lines. The stock will normally be on the upper line immediately after a cargo has been unloaded and will descend to the lower line by the time that the next cargo is ready to be unloaded.

The method described above applies to a terminal which receives a fixed quantity of LNG per year; more complex calculations must be carried out if annual receipts of LNG are increased from time to time.

Determination of Seasonal Stock

The seasonal storage level required depends more on the shape of the send-out graph than on the magnitude of the load factor, as the following simple examples show:

Example 1

The peak send-out of a terminal is 100 units/day while the daily load factor is 85 per cent, so the average send-out is 85 units/day. During the peak send-out season, the terminal receives a cargo of 180 units on alternate days, so the average replenishment rate is 90 units/day. Thus, if the send-out on any day exceeds 90 units, the excess over this figure must be drawn from the seasonal stock.

If the send-out pattern of the terminal were:

    2 days at 100 units/day
    82 days at 67.5 units/day
    281 days at 90 units/day

there would be two days with a send-out exceeding the replenishment rate of 90 so the seasonal stock would need to be:

2 days at (100 – 90) units/day = 20 units.

Example 2

If the send-out pattern of the terminal were:

    100 days at 100 units/day
    100 days at 95 units/day
    65 days at 85 units/day
    100 days at 60 units/day

the total send-out would be 31,025 units over 365 days and the average send-out would still be 85 units/day. The average replenishment rate during the peak period would still be 90 units/day, but the send-out would exceed the replenishment rate in the peak season by the following amount:

    100 days at (100 – 90)   units/day =   1,000 units
    100 days at (95 – 90)   units/day =   500 units
                  Total   1,500 units

Figure 9.6 **Calculation of Seasonal Stock Requirements**

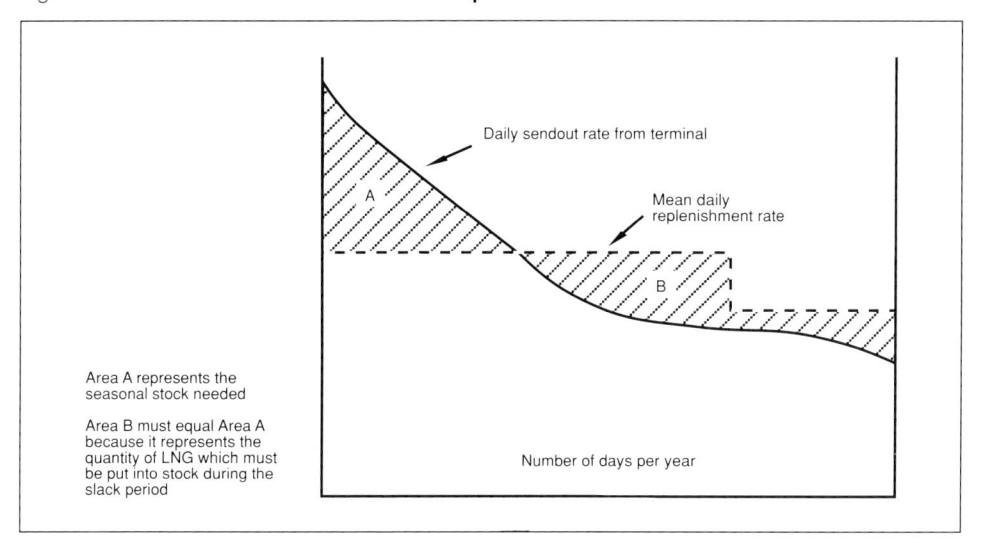

[Figure 9.6: A graph showing daily sendout rate from terminal declining over the number of days per year, with mean daily replenishment rate shown as a horizontal dashed line. Area A (above the replenishment line, under the sendout curve) represents the seasonal stock needed. Area B (below the sendout curve, above the replenishment line, on the right side) must equal Area A because it represents the quantity of LNG which must be put into stock during the slack period.]

Thus, the seasonal stock level required in this example would be 1,500 units, compared with 20 units in the first example, although the daily load factor was 85 per cent in each case.

This is illustrated in Figure 9.6 which shows that the seasonal stock level can be calculated by measuring the area between the replenishment and the send-out rates during the peak season (area A).

Table 9.1 and Figure 9.5 show that the total storage requirements can be quite large and hence costly. If underground storage is conveniently available, it may be cheaper to vaporise LNG at a fairly constant rate throughout the year and store the gas underground, thereby reducing the number of LNG tanks required.

Working Capital

The average volume of LNG at the end of a winter of average severity will be the minimum stock (average year), plus the buffer stock, plus half of the replenishment quantity. The value of this total volume is the average working capital which is tied up permanently in stocks, while the seasonal stock represents the average amount of working capital which is tied up for part of each year. Working capital requirements can be quite substantial at the higher range of LNG prices.

Simulation Models

The terminal designer must select the storage capacity, the number of berths and the capacity of the unloading lines in such a way that he minimises delays to ships without wasting money on unnecessarily large facilities.

Optimisation is facilitated by computerised simulation models which can analyse the stochastic nature of the occurrence of events such as:

- *variations in gas availability*
- *liquefaction plant maintenance and breakdowns*
- *ship dry dockings, breakdowns and weather delays*
- *delays in the loading and discharging ports caused by weather, darkness, tidal restrictions, congestion, etc.*

For example, Shell has a model which simulates the behaviour of the complete LNG project and enables the designer to see the effects of varying the storage capacity, the number of berths, etc. The designer can try various values of any design parameter and discover the effect on the number of cargoes delivered, and on the delays to ships. The model can also indicate the likely variation of performance from year to year and from season to season. The designer can use the results of computer runs to help him select the combination of design parameters which will lead to optimum performance at the lowest investment and unit operating costs.

To conclude, the owner and operator of one of the world's largest terminals has these priorities to offer:

1. Safety and reliability
2. Harmony with the environment
3. Economy including the effective utilisation of land.

Chapter 10

TRANSMISSION SYSTEMS AND DISTRIBUTION GRIDS

Historical Background

Pipelines are literally the backbone of the gas industry. Over 95 per cent of the natural gas consumed in the world today is delivered through pipes which directly link the end-consumer to the wellhead. Even with LNG, the balance of natural gas consumption, once it is regasified it can only be delivered to consumers by and through a pipeline system.

It is generally recognised that the London and Westminster Gas Light and Coke Company was the first commercial enterprise to use pipes to distribute (manufactured) gas for lighting and other purposes. This was in 1812 and the pipes were made of wood. As the use of gas spread during the nineteenth century, wooden pipes were replaced with pipes made of copper, lead, cast iron and wrought iron, until Mannesmann of Germany developed a process for the manufacture of seamless steel pipes.

By the 1920s, high-tensile steel pipe had become available and techniques were developed to construct, weld and lay large diameter, high pressure pipelines over long distances. The world's first long-distance, all-welded natural gas pipeline was laid by the Magnolia Gas Company of Dallas in 1925. This 217 mile line from northern Louisiana to Beaumont, Texas, comprised pipe of 14, 16 and 18 inch diameter. By 1955 the total mileage of large diameter, high pressure, gas transmission pipelines in the USA amounted to over 145000, and by 1990 to some 270000. The development of pipelines with these characteristics has been the single most important technological factor in promoting the growth of the natural gas industry throughout the world.

System Categories and Characteristics

Pipeline systems can be broadly divided into three main groups, namely main transmission (or trunk line) systems, regional transmission systems and distribution grids.

The function of a TRANSMISSION SYSTEM is to ensure an uninterrupted supply of gas at the desired flow rate and pressure from the point of production to the delivery point. The delivery point will often be the inlet to a regional transmission system but can also be a large volume user, such as a power plant, chemical plant, etc.

Typically, transmission pressures can be in the range of 40 to 70 bar (67 bar is approximately equivalent to 1000 psi) and pipe diameters from 60 to 120 cm (say 24 to 48 inch). Exceptionally, for example the transmission lines from West Siberia to Europe, such lines can be as long as 5000 km with diameters up to 140 cm (56 inch). Length is an important design consideration as a stream of gas in a pipe is subject to friction. As the gas flows, so its pressure drops. Double the length of the pipeline and the friction is doubled, so the pressure drop is greater. At peak periods of demand when the flow rate is increased, the pressure drop also increases.

To ensure that gas is received at the delivery points at the desired rate of flow and at the required pressure, the designer of the system can consider either laying a pipeline(s) parallel to the original line, known as 'looping', and/or increasing capacity by adding compressor station(s) which boost the flow and pressure in the line. In general, a pipeline requires greater investment than a compressor station but operating costs are lower, and as most compressors are gas-fuelled, delivered quantities of gas are correspondingly reduced. The designer of the system has to balance out these capital and operating costs differences to arrive at an optimum solution. However, if demand continues to increase, there will come a point when looping becomes inevitable.

At the delivery point the gas is metered and its pressure reduced for acceptance by the regional transmission system. At this point natural gas, which has virtually no smell of its own, is odourised to facilitate the detection of leaks.

The function of a more localised REGIONAL TRANSMISSION SYSTEM is to take gas at the delivery points of the main transmission system and transport it to local gas transfer stations sited at or close to the main areas of consumption. This transfer point is often known as the 'city gate'. Here the gas is metered again and the pressure further reduced for acceptance by the distribution grid.

Typically, regional transmission pressures can be in the range of 15 to 40 bar and pipe diameters up to 60 cm. Increasingly, regional systems are laid in rings around conurbations to enhance their reliability. Thus in the event of a section having to be shut down for any reason, the flow of gas can be reversed and supply maintained.

In some countries, regional transmission systems in terms of their ownership and operatorship are an integral part of the main transmission system, in others they may be part of the distribution grid. They are not discussed any further as it is considered that for the purposes of this book they are covered sufficiently by the commentaries on transmission systems and distribution grids.

DISTRIBUTION GRIDS consist of a complex network of supply mains, feeder mains, distribution (or street) mains and services (small pipes) which deliver the gas from street mains to customers' meters.

From a pressure of about 10 bar or less at the outlet of the city gate station, pressure is progessively reduced at various points in the network down to around 25 mbar before it is delivered to residential and other small-scale users. As touched on earlier, large volume users, such as major industries, chemical plants, power stations, etc., may be supplied directly from main transmission or regional pipelines, sometimes directly by producers.

At the city gate station gas is metered and odourised if this has not been done at an earlier stage. It may also be passed through a cleaner to remove any entrained liquids and dust.

Some design considerations and the need to provide supplementary forms of gas supply to cope with periods of high demand are discussed elsewhere.

Figure 10.1 **A Metering Station, Epe, Germany**

Courtesy of Ruhrgas A.G.

Practical Aspects of Overland Transmission Pipelines

Before a cross-country transmission pipeline can be constructed, the route has to be selected and surveyed and plotted on maps, rights-of-way along the proposed route have to be negotiated with landowners and permits sought from the relevant authorities. This is a time-consuming process, particularly if the land to be traversed is congested with roads, railways, waterways, buildings and the like, and/or is of an environmentally sensitive nature or, yet again, if land is owned in many small, separate tracts.

While these problems are being tackled, detailed design of the pipeline can proceed. As mentioned earlier, the main factors to be taken into account are the volume to be transported, the rate of flow and pressure required, and the distance involved. Other equally important considerations include the nature of the terrain to be crossed – rocky, swampy, wooded, waterways, earthquake prone, etc. – the type of corrosion protection needed, provision at the outset or later for compressor stations, control systems, and possible connections to other pipelines and storage installations.

The right-of-way is a strip of land usually from 10 to 30 metres wide depending upon the type of terrain and size of pipe to be laid. Once the go-ahead is given, this

Figure 10.2 **A Spread in Action in Scotland**

Laying the 220 km, 16 inch gas liquids pipeline from St. Fergus to Mossmorran.

Figure 10.3 **Coating and Wrapping Pipeline Welds in The Netherlands**

Courtesy of N.V. Nederlandse Gasunie.

strip of land will need to be cleared of all trees and undergrowth, good topsoil removed and stockpiled for eventual replacement, rocky outcrops blasted or ripped away, fences removed and resited, preparations made for road, rail and water crossings, etc. Sites also have to be found where pipes can be stocked, cleaned, wrapped and coated before they are laid.

These preparatory activities, as well as the ditching and the laying of the pipe itself, are undertaken by 'spreads'. A spread usually comprises several hundred men and all the equipment involved in building a pipeline from untouched land until the line is laid and the ground is restored as near as is practicable to its original state. Under favourable terrain conditions a spread may stretch over a strip of land of up to 20 kilometers in length and lay pipe at the rate of up to 3 kilometres per day. For long pipelines, several spreads will be employed, leap-frogging each other as they complete their sections. Obviously, in more difficult conditions, spread coverage will be much shorter and laying rates much longer.

Once the surface of the right-of-way strip is cleared and graded, a continuous track capable of supporting the ditching machines and sideboom tractors is constructed parallel to the staked line of the surveyed pipe route. Ditching machines then scoop

Figure 10.4 **Laying a Pipeline Across a River**

A pipeline about to be laid in a trench across the River Dee, Scotland.

out ditches or trenches deep enough to allow the surface to be cultivated without damaging the pipe once the pipe is laid and the trench is backfilled. The width of the ditch will be kept as narrow as possible but sufficient to avoid any damage to the protective coating of the pipe as it is lowered in. In rocky terrain ditching can involve drilling, blasting and ripping.

At the pipe stock yard, pipes are cleaned externally, primed, coated and wrapped with appropriate corrosion protection materials before they are moved onsite where they are welded together. Clearly, first class standards of welding, manual or automatic, are essential. All welds will be inspected visually and a percentage by radiographic and/or ultrasonic means, except where water, road and rail crossings are involved when it is customary practice to X-ray all welds. Welds are then cleaned, coated and wrapped. Detectors are used to check for any flaws in the coating.

Finally, bending machines, which can bend pipe laterally and vertically, bend the welded pipes to align with the contours of the ditch before the pipeline is lowered by sideboom tractors into the ditch, which is then backfilled and made good.

Somewhat different procedures are involved for wide or deep water crossings. In such cases the pipes are usually made of thicker steel and welded together onland. A reinforced concrete coating is applied to give the pipe mechanical protection and negative buoyancy before it is pulled by tugs or winched into place. Alternatively, the pipeline may be carried across the water on a specially constructed span bridge. To avoid disrupting traffic, tunnels are usually bored under roads and railways through which the pipe can be pushed or pulled through. Again thicker steel is used at such points. In marshy or swampy conditions concrete weights or anchors may be needed to keep the pipe from floating, while in permafrost areas pipes are sometimes installed on stilts above ground to avoid the pipe sinking as it warms up and melts the permafrost as gas passes through it.

Pipeline laying is now so commonplace that it may be regarded by some as a very straightforward and relatively simple activity. In practice, depending upon the nature of the terrain, it can be a very complicated and demanding task requiring

Figure 10.5 **Lay Barge 'ETPM 1601' in Action Offshore Western Australia**

Laying pipe from the North Rankin field to shore at the rate of about 2 km per day. The stinger can be seen in the right foreground. Courtesy of Woodside Offshore Petroleum Pty. Ltd.

Figure 10.6 **A Reel Laying Ship in the North Sea**

considerable engineering skills and sophisticated logistical planning and control techniques.

Because of the substantial differences in the terrains encountered, labour costs, steel prices and rights-of-way payments, rules of thumb for the cost of constructing a pipeline would be misleading.

Practical Aspects of Subsea Pipelines

Increasingly, gas pipelines are being laid in deep water as the search for gas extends offshore into deeper waters. Techniques have been developed which enable transmission and gas gathering lines to be laid successfully in water depths exceeding 1000 metres.

Most offshore pipelines of any length are laid by specially designed lay barges on which the pipe is cleaned, primed, coated, wrapped, welded and inspected onboard. Submarine pipelines are also coated with reinforced concrete to afford them mechanical protection and negative buoyancy.

Modern lay barges are usually self-propelled, equipped with dynamic positioning devices and all facilities to house several shifts of workers so that they can operate

Figure 10.7 **Submarine Pipeline Plough**

This 380 tonne plough was designed for the N.W. Shelf project. The plough trenched to a depth of one metre in tough limestone and to 2 metres in softer seabed material. Courtesy of Woodside Offshore Petroleum Pty. Ltd.

continuously, 24 hours a day, even during severe weather conditions – see Figure 10.5. Once the pipe is ready for laying, it is lowered into the sea via a supporting arm or 'stinger' which can be adjustable down to about 60 degrees from the horizontal. Alternatively, for smaller diameter lines, flexible pipes may be used and fed into the sea via a reel or drum – see Figure 10.6.

Depending upon the sea bed conditions, water depth, tidal and sea traffic movements, submarine pipelines may rest on the sea bed, or placed in a trench gouged by a bury sled or plough (see Figure 10.7), or placed on the bottom and covered by deposited rocks. Underwater TV cameras and divers help to ensure that the line is correctly positioned.

Two Phase and Dense Phase Flow

Apart from the obvious difference in the environment, the other major difference between onland and submarine pipelines is that in the former case the gas to be transported is usually of marketable quality, whereas submarine lines normally transport gas from production platform to shore where it is processed to the required quality. This is because it is not normally economic and/or practical to install sophisticated gas processing facilities on offshore platforms. If the gas as produced is relatively free of liquids ('dry gas'), then piping to shore is relatively straightforward. If, however, it contains a high proportion of liquids ('wet gas') then this presents a challenge to the pipeline designer.

Figure 10.8 **A Pipeline Pig**

This pig is used to sweep the 450 km, 36 inch pipeline from the Brent field to St. Fergus.

There are two principal methods of piping 'wet' gas; they are two phase flow and dense phase (high pressure) flow. Two phase flow is usually preferred where the volumes of gas to be transported are small, distances are fairly short and where the liquid content is not too high. The disadvantage is that the line may need regular and frequent 'pigging' to remove the liquid slugs as they collect in the pipeline. This is costly and reduces the volume of gas that can be transported through a pipeline of a given diameter. Other difficulties include the inability to predict accurately the

Figure 10.9 **Typical Phase Envelope for a Natural Gas Mixture**

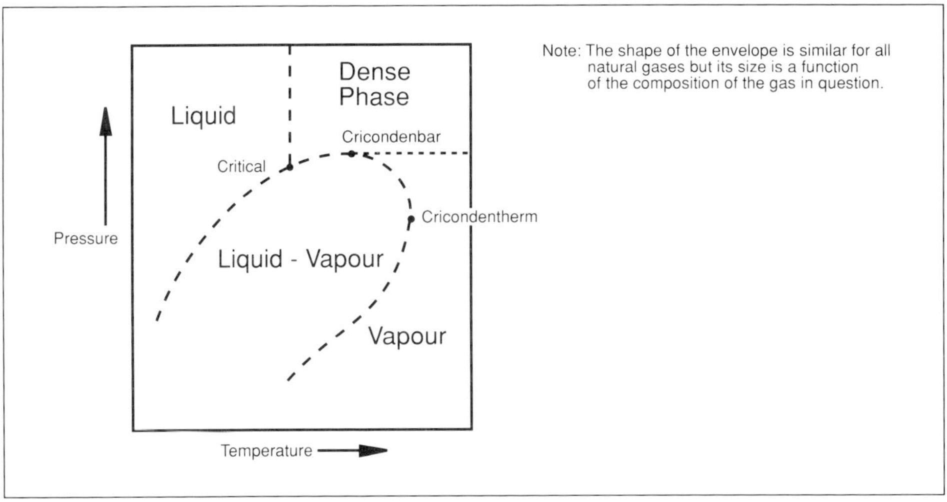

Note: The shape of the envelope is similar for all natural gases but its size is a function of the composition of the gas in question.

behaviour of two phase flow with the result that liquid slugs arrive at the reception terminal erratically and are of different sizes.

Dense phase flow, a more recent development, operates in a single phase neither liquid nor vapour, which avoids some of the difficulties of the two phase flow, but requires the minimum operating pressure of the pipeline to be greater than the cricondenbar of the gas being transported. Cricondenbar is the highest pressure at which separate vapour and liquid can exist. Depending on the composition of the gas, this can be in excess of 100 bar. Such pipelines have to operate at temperatures higher than the cricondentherm of the gas being transported. Cricondentherm is the highest temperature at which liquid can form, regardless of pressure – see Figure 10.9.

For conventional gas pipeline operations the gas to be transported must have a cricondentherm below the lowest expected operating temperature. For a subsea pipeline this can be below likely minimum water temperatures due to the cooling effect as gas expansion occurs as a result of pipeline pressure drops.

The technology, materials and equipment now exist to construct and lay large diameter, high pressure pipelines offshore. Both two phase and dense phase pipelines have their economic and capital and operating costs advantages and

Figure 10.10 **Waidhaus Compressor Station, Germany**

Courtesy of Ruhrgas A.G.

disadvantages and these can only be determined in relation to the specific circumstances and conditions involved.

Compressor Stations

For long distance pipelines compressor stations are essential in order to raise the pressure lost by friction as the gas flows through the line back up to its required level. Compression, within limits set by the maximum operating pressure of the pipe, is also an alternative to building additional pipeline capacity. The amount of compression required will be a function of the diameter and length of the pipeline, the desired operating pressure, and the amount and rate at which gas is taken off the pipeline at its downstream delivery point(s) at any given moment. Compression will also be needed at the inlet of the pipeline once the natural pressure of the produced gas has fallen below the required pressure of the gas stream to be transported.

In simple terms, modern gas turbine compressors – there are other types – consist of a vaned motor, turning at very high speeds, which raises the velocity of the gas. This velocity is then reduced and its kinetic energy converted into pressure via a

Figure 10.11 **Interior View of The Ommen Compressor Station, The Netherlands**

Courtesy of N.V. Nederlandse Gasunie

static diffuser. Compressors are usually fuelled with a mixture of filtered air and gas taken off the pipeline. The mixture is ignited and expands through the turbine blades which then drives the compressor. Typically, pressure of the flowing gas stream is raised by a factor of about 1.3, e.g. from say 50 bar to 65 bar. This would require about 3000 m^3 of gas to drive the compressor to raise the pressure of one million m^3 of gas by 15 bar. If a higher pressure increase is required, this can be achieved by passing the gas through a second stage of compressors.

Each compressor station usually comprises several compressors working in parallel, or in series if they are of the low pressure ratio centrifugal type, with sufficient capacity to allow for periodic off-line maintenance and repair. Many stations are fully automatic but capable of being manually controlled. As compressors are inherently very noisy, careful siting and extensive sound insulation are necessary. Safety precautions such as gas leak detectors and emergency shut down equipment are essential, especially in unmanned stations controlled remotely.

Recent developments aimed at improving the efficiency of gas turbine compressors include recovering the waste heat from the exhaust gases for steam generation. The steam can then be used to produce electricity and/or for a

Figure 10.12 **Effect of Volume on Costs**

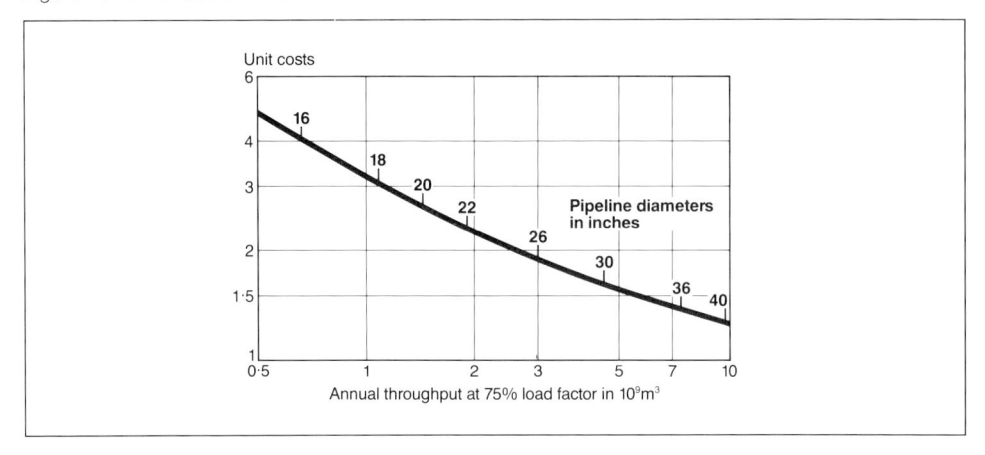

[Graph showing Unit costs vs Annual throughput at 75% load factor in $10^9 m^3$, with pipeline diameters in inches labeled: 16, 18, 20, 22, 26, 30, 36, 40]

condensing steam turbine driven centrifugal compressor. Depending upon the configuration and types of compressors used, this can increase efficiency by some 10 per cent. The heat of the exhaust gases can also be used to pre-heat the air of the air-gas fuel mixture during periods of low ambient temperatures thereby reducing fuel consumption. Environmental standards can also be enhanced by thermally treating the exhaust gases to reduce nitrogen oxides emissions.

Basic Pipeline Economics

Pipeline economics will reflect many of the design aspects mentioned previously. This section focuses on just three aspects – volume, load factor and the relationship between fixed and variable costs – in a very broad and generalised way to illustrate what is in practice a highly complex assessment.

Figure 10.13 **Effect of Load Factor on Costs**

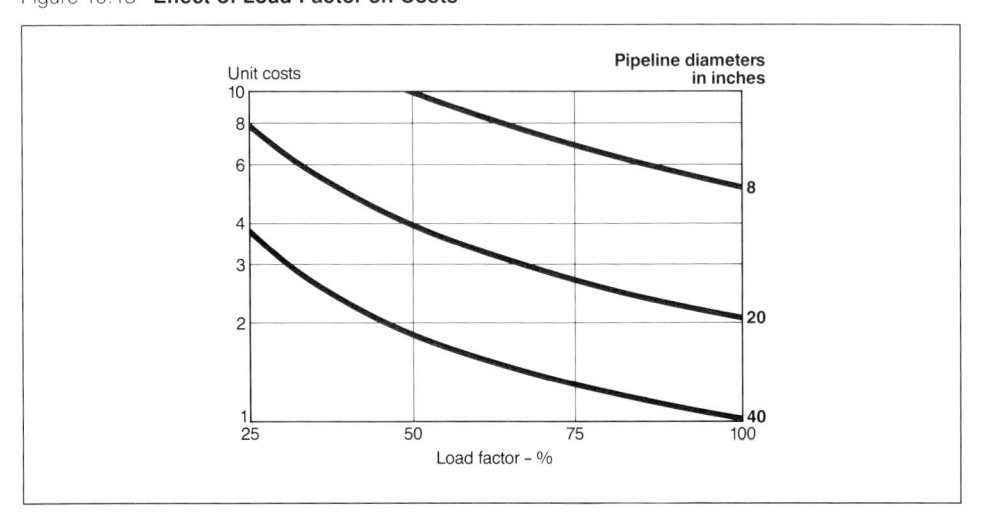

[Graph showing Unit costs vs Load factor - %, with pipeline diameters in inches labeled: 8, 20, 40]

Page 140

As far as volume is concerned, and ignoring for this purpose the added complications of distance, operating pressures, compressor stations, etc., the larger the volume of gas to be transported, the larger the diameter of pipe needed. Figure 10.12 demonstrates in schematic form that as the volume of gas to be transported increases so unit costs progressively decline assuming that other elements, including load factor (assumed to be 75 per cent), remain equal. In this example, if it requires an 18 inch diameter pipeline to transport 1 Bcm of gas per year and a 40 inch line for 10 Bcm, the latter would cost less than half the former on a unit basis.

With regard to load factor, Figure 10.13 illustrates that as the load factor improves so unit costs fall. For example, the unit costs for a 20 inch line at a 25 per cent load factor are four times greater than at 100 per cent. So in theory at least, although

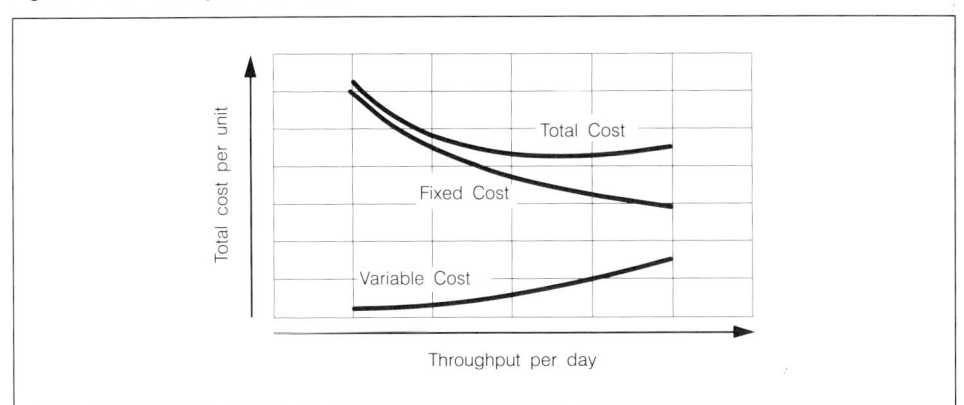

Figure 10.14 **Transportation Cost for a Particular Diameter**

there can be some inconsistencies in practice, the greater the volume and the higher the load factor, the lower the unit cost of transportation will be.

In unit terms, fixed costs are invariably greater than variable costs under virtually all operating conditions. Fixed costs are the direct function of the original investment and include depreciation, the required return on the investment, insurance, certain taxes, etc., while variable costs include labour, maintenance and repair, fuel consumption (e.g. for compressors), administration, etc. As throughput for a given diameter pipeline increases, fixed costs per unit will normally fall and variable costs will rise as respective proportions of total unit costs – see Figure 10.14. There are, however, some exceptions to this generality. For instance, fixed costs per unit will rise temporarily when the pipeline is looped and/or when compression has to be installed if there is little or no immediate increase in throughput. However, they will fall again as increases in throughput build up.

Practical Aspects of Distribution Grids
The design of a distribution grid takes account of five main activities – system design,

Figure 10.15 **Laying a Large Street Mains at Night**

Courtesy of Tokyo Gas Co. Ltd.

flow analysis, routing, materials and safety assessment. As there is seldom one obvious solution, it is an iterative process to evaluate alternative designs so as to find one which is safe, efficient, economic and feasible for the environment and terrain in question.

SYSTEM DESIGN starts with estimating the likely growth in gas demand over a specified period. This is usually the responsibility of marketing and planning specialists. Then they, together with pipeline engineers, will assess and calculate the peak instant demand which the system may have to handle under the most severe weather conditions that may occur once in say 20 years. The latter will take into account the expected customer and appliance (use) mix and weather conditions for those customers with temperature-sensitive applications primarily, of course, space heating or air conditioning.

In practice, actual peak demand will be less than the summation of the calculated individual customer's peak demand as customers' habits and appliance use varies. This phenomenon is known as 'diversity of demand' and formulae exist to calculate it.

Gas pressures at the various entry and exit points of the grid and gas velocities throughout the grid also have to be incorporated in the design parameters, and consideration given as to the desirability of incorporating peak shaving facilities to avoid over-sizing the gas mains.

FLOW ANALYSIS is an assessment of the gas flows and pressure drops in the system allowing for different pipe lengths and diameters, friction factors, any changes in altitude, average temperature and compressibility of the gas under typical flow conditions, the quality and specific gravity of the gas to be distributed, etc. Other than for a very simple system, it is a very complex calculation necessitating the use of a digital computer and a tailor-made program.

ROUTING of the grid is first plotted on large scale maps and then visually surveyed to check for possible physical obstacles and locations where pipes could be subjected to abnormal loadings, e.g. heavy vehicular traffic, subsidence or other adverse conditions. Local authorities will need to be consulted on forthcoming building works and planning applications. Alternative routes may then need to be plotted and surveyed. Drawings will be required for certain types of crossings, e.g. waterways, major roads, railways, etc., to show construction details. Final plans and maps of the route are then prepared, noting where wayleave arrangements and other special agreements will need to be negotiated.

PIPE MATERIALS may be welded steel, mechanically jointed ductile iron and fusion jointed polyethylene, or most likely a mix thereof, i.e. that material which best meets the nature and duty of the pipe at the location in question. Selection will also be influenced by maximum operating pressures and flow rates, minimum and maximum operating temperatures, the degree of impact and corrosion resistance needed, the nature of the environment, etc. Isolating valves, pressure regulation valves, filters and other equipment will need to be selected and ordered.

The primary objective of SAFETY ASSESSMENT is to avoid the possibility of an uncontrolled release of significant quantities of gas in the vicinity of people and buildings. This will affect the routing of the system, the materials and equipment used, methods of construction, the degree of corrosion protection to be afforded and the need for special precautions at potentially hazardous locations.

The probabilities of gas entering buildings and collecting in other potentially dangerous areas in the event of a mains failure have to be assessed and catered for to ensure that the design of the grid has the necessary integrity appropriate to the environment.

Only when all these matters have been thoroughly investigated and determined, and all necessary permissions from local authorities obtained and wayleaves granted, can construction commence.

Construction and Commissioning

Key considerations during construction work are ensuring the highest possible safety standards are maintained at all times, minimising disturbance to the general public, ensuring there is no damage or disruption to other essential services, and ensuring that construction sites are left in at least as good a condition as when work started.

When the grid, or sections of it are completed, they will be hydrostatically tested for possible leaks, purged with an inert gas and then with natural gas which will be vented and burnt before commissioning.

Finally, the obvious is often overlooked or taken for granted. This is particularly so in the case of the supply of gas by pipeline which by its very nature has no apparent visible manifestation. The average consumer seldom appreciates that he, or she, can be directly connected by a continuous pipe to the bottom of a well in some gasfield which may be several thousand kilometres away and that the pipe may have crossed several countries to reach him.

Chapter 11

THE RESIDENTIAL MARKET

Evolution of the Market

Low calorific value manufactured gas was first used for residential purposes in the early years of the nineteenth century, principally for home lighting. In those days gas was sold on a 'rental' basis according to the number and size of burners (lights) and hours they were used. Subsequently, in the 1830s, following the invention of the gas meter, gas was increasingly sold by volume. During the mid to late 1800s other inventions such as gas-fired cookers, water heaters, room heaters, soldering irons, hair-curling tongs and many other ancillary appliances, led to a much wider use of gas for domestic purposes. Nevertheless, lighting continued to be the largest application in most countries until shortly before the first World War.

The demise of gas lighting was brought about by the invention of the electric dynamo and the subsequent development and availability of electricity for general lighting (and power) purposes. By about 1920, few homes continued to be lit by gas. The wheel has now turned a full circle, and with the development of decorative outdoor natural gas lights for gardens, patios, driveways, residential streets and so on, more natural gas is now consumed in the United States for such purposes than was the case at the height of the manufactured gas light era 80 years or so ago.

The importance of the use of ('wet') manufactured gas for lighting was that it created a vast gas distribution infrastructure. Thus when natural gas became increasingly available and displaced manufactured gas, there existed the means to supply natural gas to many millions of households in North America, Europe, the Soviet Union, and in a number of countries in the Far East and South America as well. Admittedly, many of these distribution systems required reinforcement, partial replacement or modification to accept and distribute high calorific value ('dry') natural gas safely and efficiently, but the basic structure was largely in-place. And it was the arrival of natural gas that resulted in the rapid expansion of the use of gas for cooking, water heating, space heating and other applications for which manufactured gas had certain limitations in terms of efficiency, safety or cost compared with oil and electricity.

Conversion

The distinction made above between 'wet' manufactured gas and 'dry' natural gas is simply to highlight that the introduction of natural gas involved much more than just a change in gas quality. Many manufactured gas distribution systems were constructed with cast iron pipes which were not welded together. Pipe joints were packed with hemp or some similar fibrous material to achieve gas tightness and were kept moist and in place by the 'oily' content of manufactured gas. These joints dried out and leaked when natural gas was introduced, necessitating installing leak-proof joint clamps and other methods to ensure that the pipes were gas-tight and safe.

Customers' appliances also had to be converted in order to operate satisfactorily and safely with the higher calorific value of natural gas. In most countries conversion was carried out at no cost to the consumer, except when it was too difficult to convert an old appliance in which event customers were usually offered very generous discounts on new ones.

As may be imagined, conversion was a much more complex, time-consuming and costly undertaking than the foregoing may imply.

Market Characteristics
No particular residential gas market can be regarded as being representative or typical. Each market has its own unique characteristics and will reflect and be conditioned by such factors as:

- *the eating, living and social habits and customs of the local population;*
- *the various types of dwelling, their methods of construction and materials commonly used;*
- *average disposable incomes and how they are spent;*
- *standards of comfort desired;*
- *the availability and cost of competitive fuels; and,*
- *variations in and the severity or otherwise of ambient temperatures.*

It is a huge market for gas. For example, in round numbers there are 50 million residential gas consumers in the United States, 17 million in the UK, 9 million in (West) Germany, $5\,^1/_2$ million in the Netherlands, 9 million in France and 21 million in Japan. In a number of countries, the residential sector is the largest individual market sector in volumetric terms.

As discussed elsewhere, more often than not residential gas prices are controlled or regulated by local or national authorities and, once connected, local distribution companies usually have a legal obligation to supply their customers at all times provided they pay their bills.

In the majority of well-established gas markets in the northern hemisphere, the main residential uses are space heating, water heating and cooking, in that order of importance. A selection of some of the appliances used for these purposes, new developments, details of competitive fuels and other data are given in subsequent sections. In specific locations, clothes drying, refrigeration, the heating of swimming pools, air conditioning, barbecues, decorative lighting, etc., are also worthwhile outlets for gas.

Cooking

The three main gas consuming cooking appliances are hot plates, grills and ovens; rice boilers are important in Far Eastern markets. Although annual gas consumptions per connection for cooking purposes are rather small, the cooking flame in the kitchen is the only visible manifestation of gas for most consumers. Moreover, since the earlier days of gas lighting, cooking can be the main reason why a dwelling has a gas connection in the first place.

In the modern house, electricity is the main competitor to gas for cooking. Because annual consumption per connection of either energy is relatively small energy costs play little, if any, part in their selection. Choice may be dictated by what the housewife was brought up on and accustomed to use, or by the attractiveness and cost of the appliance. Electricity has a good modern image. Electric appliances are often cheaper, available in a wider choice, are thought to be easier to clean and produce no products of combustion at the point of use. Gas on the other hand responds faster, gives an easy visible check on whether the burner is on, and provides a much better heat transfer to non-flat bottom utensils. Many cooks prefer gas for these and other reasons.

Cooking has become a highly competitive section of the market which the gas industry, after many years of largely having its own way, was rather slow to recognise. In more recent years, the gas industry has fought back strongly and has developed more attractive and efficient appliances using electricity, e.g. for electronic ignition instead of pilot flames, time switches in ovens, etc., where it suits its purposes. Easier to clean, smooth top cookers are now available, as are burners which produce less NO_x. These are just some examples of the new developments being introduced.

Water Heating

The choice of water temperature depends on use, typically about 40°C for hand-washing, showering and bathing, around 50°C for dishwashing by hand and 65°C by machine, and up to 80°C for clothes-washing. As many dish and clothes washers can operate on cold water, water heating appliances that produce water temperatures in the range of 40 to 60°C suffice for most residential purposes.

If the dwelling is equipped with a suitably designed water-based space heating system (see later), then this is usually the most convenient and economic way of providing all the hot water needs of the household. If not, hot water can be provided by either gas-fired instantaneous water heaters or hot water accumulators. Electric powered heaters and accumulators are the main competitors to gas.

Unflued instantaneous water heaters installed over a sink or a bath or in a shower unit, are the easiest and cheapest means of providing a continuous flow of hot water at the point of use. Their disadvantages are that the hot water flow rate is fixed, and in confined unventilated locations products of combustion can present a safety hazard. The first drawback can be largely overcome with burners that automatically modulate with the hot water off-take rate, but they are more expensive and less efficient. The second can be resolved either by flueing the heater to the outside atmosphere, or by installing it outside the dwelling, but both options increase the cost of installation and may not always be practicable. These costs can be partly offset if the appliance is equipped with electronic ignition instead of a gas pilot. Despite some potential disadvantages, instantaneous water heaters are nevertheless very popular, and provided they are properly installed and maintained, present no safety hazard.

Hot water accumulators are generally preferred over instantaneous water heaters in North America and some European countries. Their efficiency has improved considerably in recent years by using better insulation methods to reduce heat losses, by replacing gas pilots with electronic ignition, by storing hot water nearer to its required temperature and by improvements in burner and flue designs. Efficiencies of over 90 per cent have been achieved with immersed pulse-combustion heaters.

Space Heating

Space heating, in particular central heating as distinct from local (single room) heaters, is the major residential use of gas in volumetric terms in most temperate and cold climate gas markets. The focus of this section is on central heating, although flued and unflued (where permitted by the authorities) local heaters are also an important outlet for gas in many countries. It is appropriate to note at this point that for safety reasons central heating boilers and other large capacity appliances have to be flued to outside atmosphere if installed inside the dwelling. Figure 11.1 illustrates two typical balanced flue arrangements, also the so-called Modified SE Duct for multi-flueing in an apartment block. There are other flueing concepts.

Traditionally, in most European countries water is the main medium for central heating, whereas it is air in North America. Water has the advantages of better heat transfer, it requires narrow tubing rather than wide ducts for air circulation and is quieter, but radiators can sometimes be obtrusive. A further important advantage is that water-based central heating can be combined easily with water heating for other purposes.

The advantages of air are simplicity, particularly if the ducts are installed when the dwelling is being constructed, no radiators, no pumps, the ease with which it can be combined with ventilation or air conditioning, and the low inertia of the system.

Figure 11.1 **Typical Flueing Arrangements**

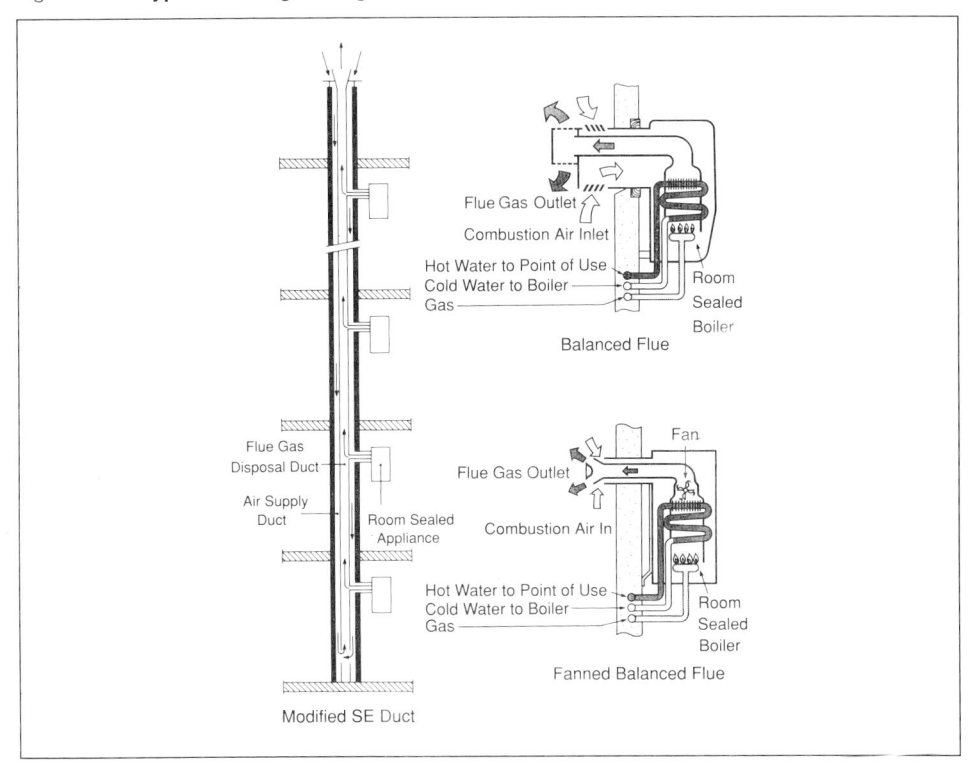

It readily lends itself to the use of heat pumps which are discussed later. Outside North America, water-based central heating is still favoured, particularly for installing in existing dwellings, although air systems appear to be gaining in popularity.

The main competitors to gas-fired space heating systems are usually domestic heating oil and electricity. However, in some countries non-gas fuelled district heating systems, coal, LPG and, notably in Japan, kerosene, are also important competitors to gas. All these alternatives have their various advantages and disadvantages compared with gas. These can and do vary considerably from country to country and thus are very difficult to quantify in general terms. For example, electricity can be much more expensive than gas to use at the full tariff rate, but not necessarily so at off-peak rates in combination with storage heaters in low-energy designed dwellings. Domestic heating oil prices vary so widely, as do gas prices for that matter, as to make it impossible to generalise on their relative cost competitiveness. However, oil systems require a bulk storage tank which can be a disadvantage where available space is limited, and there can be corrosion problems in flues and heat exchangers with certain types of boilers arising from oil's sulphur content.

Comparisons are not helped by the fact that in some instances the capital cost of electricity, oil and gas-fired systems may be subsidised in some way by the utility or oil company concerned to help 'capture' the outlet. A further complication is that for most new dwellings the decision taker as to which fuel to use for heating purposes is not the eventual householder, but the architect, builder, property owner or local authority. Their decision may not always be based on the relative efficiencies and operating costs of the options open to them, but more on what they think the householder may prefer, or because investment rather than operating costs are their main criterion or, yet again, because the prospective fuel supplier offers some incentive. Equally, some systems may not be practicable to install in existing dwellings not previously equipped with central heating.

Obviously, selection of the fuel and system to use is not as straightforward as it might first seem, the more so since most householders do not have the expertise to make precise comparisons and are open to persuasion by sales talk and hearsay.

Unlike the industrial market, the price elasticity of demand (PED) of the residential space heating market is, within reasonable limits, relatively inelastic. PED can be defined by the following equation:

$$PED = \frac{\text{Percentage change in quantity demanded}}{\text{Percentage change in price}}$$

It is inelastic because once the chosen central heating system is installed the average householder is unlikely to change his installation for 10, 15 years or more, i.e. until such time as replacement becomes necessary. If during that time the price of his particular fuel rises much more than the rate of inflation and/or the price(s) of competitive fuels, the householder is more likely to improve the insulation of his dwelling and/or control his comfort level more carefully to reduce his fuel consumption and running costs. Replacement of an existing reliable boiler with a more efficient version, or with a different and cheaper fuel system, is seldom the first choice of the majority, unless some inducement which is perceived to be worthwhile is offered.

Paradoxically, as gas and electric utilities, oil distributors and other energy suppliers compete to gain an increased share of the lucrative space heating market, average annual consumptions per household are, in general, falling when corrected for ambient temperature variations. This is because of the much greater efficiency of modern appliances, improved home insulation, smaller sized and better designed dwellings and, in many developed countries, fewer persons per family as birth rates decline. Nevertheless, securement of the space heating business continues to be the prime target of the residential market for energy suppliers. In this regard, for high efficiency dwellings the need to eliminate fixed costs, i.e. gas's so-called 'standing charge', is becoming an increasingly important consideration.

In the following sections some of the more recent developments in gas-fired central heating appliances are described.

Condensing Boilers

Conventional non-condensing water-based central heating boilers are being progressively superseded by condensing boilers. However, non-condensing boilers still predominate and may be the only cost-effective replacement boiler if the rest of the heating system requires modification for satisfactory performance with a condensing boiler.

Combustion of natural gas produces mainly water vapour and carbon dioxide, viz:

$$CH_4 + 2O_2 \rightarrow CO_2 + 2H_2O$$

Virtually all the water vapour and carbon dioxide produced, and any other products of combustion, e.g. nitrogen oxides, travel up the chimney or flue and are

Figure 11.2 **Schematic Cross Section of a Condensing Boiler**

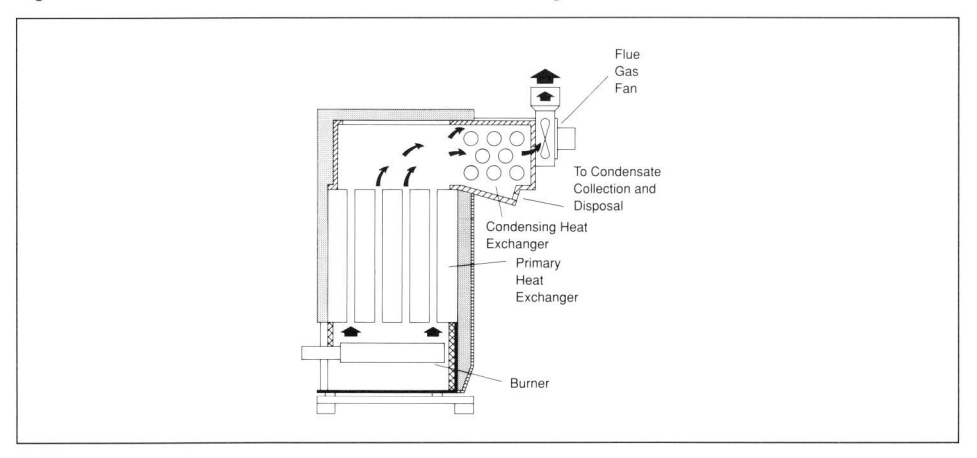

expelled to atmosphere as vapours. However, if all or most of the generated water vapour can be condensed, efficiency is improved by recovering the latent heat of condensation which would otherwise go up the chimney.

In practice, and as illustrated in Figure 11.2, the principle of a condensing boiler is that the cooled water returning from the radiators flows first through the condensing heat exchanger, picking up heat as it does, then through the primary heat exchanger before being pumped back to the radiators.

Condensing boilers must have fanned flues, which have the advantage of negligible stand-by losses, because the low temperature of the combustion products reduces the natural flue draught; most are equipped with electronic ignition instead of pilot flames. Boilers of this type can result in a gas saving of about one-quarter over a conventional modern boiler of a similar capacity for a similar duty.

Figure 11.3 **Hydro Pulse Condensing Boiler and Principle of Pulse Combustion**

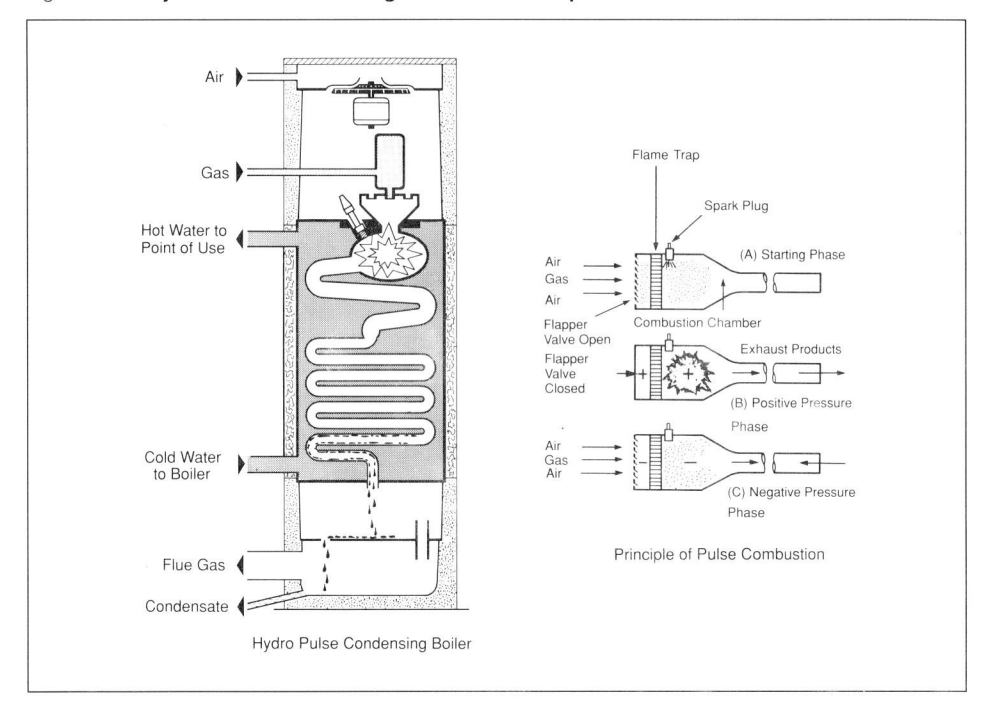

A variation of the above is a condensing boiler with pulse combustion, the principle of which is illustrated in Figure 11.3.

Once the combustion process is set in motion with a start-up fan and an ignitor, it is self sustaining. Although the pressure in the combustion chamber ranges from below to above atmospheric, its mean value is positive and high enough to overcome the pressure drops caused by the heat exchangers and vent pipes. The main disadvantage of pulse combustion is noise which necessitates installing the appliance outside or the provision of sound insulation.

Heat Pumps

In simple terms, heat pumps transfer heat from low to high temperatures. When heat pumps cool, e.g. as in refrigerators, they withdraw heat at low temperature from inside the refrigerator and reject it at a relatively higher temperature into the kitchen. When heat pumps are used for space heating they perform the same operation, only at higher temperature levels, withdrawing heat from the ambient air or ground water outside a building and supplying heat at higher temperatures inside the building. Their principal merit is that they use ambient air or water as an inexhaustible and 'free' source of heat or cold.

A heat pump comprises three main components through which a fluid is circulated. They are an evaporator where the fluid absorbs heat by evaporating, a

Figure 11.4 **Heat Pump Concept**

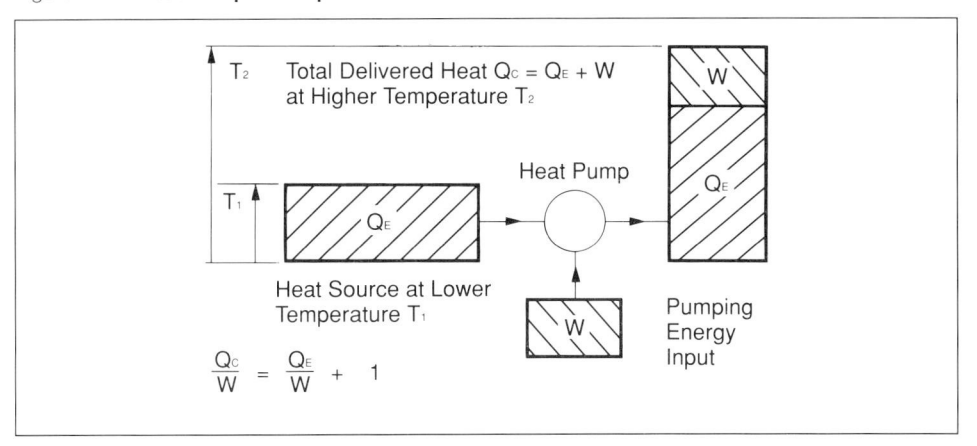

[Figure: Heat pump diagram showing heat source at lower temperature T_1 with Q_E, heat pump with pumping energy input W, and total delivered heat $Q_C = Q_E + W$ at higher temperature T_2. Equation shown: $\frac{Q_C}{W} = \frac{Q_E}{W} + 1$]

condenser where the fluid rejects heat by reverting to its liquid state, and a 'motor', typically a compressor, to keep the fluid circulating. The gas application is, of course, providing the energy to run the motor. Figure 11.4 illustrates the heat pump concept.

Where:
Q_c = the amount of heat picked up (T_1)
Q_e = the amount of heat rejected by the pump (T_2)
W = the energy used to work the motor

Efficiencies are very sensitive to T_2 minus T_1, and they can be very high where the required temperature lift is small. However, when the ambient air (heat source) gets colder, the heat pump produces less heat. Accordingly, in cold climates heat pumps have to be oversized and usually complemented by a conventional heater which makes the heat pump more expensive and less efficient. But if nearby ground water is available, which is not generally the case, efficiency is likely to be higher than an air-sourced pump.

Heat pumps are gaining in popularity for both residential and commercial uses as technological developments improve their efficiency. They are much more practical when they are designed and can be used for air cooling in the summer and space heating in the winter. The main competitor of the gas-fuelled heat pump is the electric heat pump which currently dominates the market. Gas heat pumps suffer from the low mechanical efficiency of gas engines compared with electric motors, but the waste heat makes them more attractive in colder climates. Economics in general depend on local climate and the electricity: gas price ratio.

Chapter 12

THE COMMERCIAL AND TRANSPORTATION MARKETS

The commercial (gas) market can be defined as that group of end-consumers who are engaged primarily in service type activities, i.e. shops, supermarkets, hotels, restaurants, schools, hospitals, offices, etc. These outlets can vary considerably in their gas usage from a small shop to a large city hospital or hotel complex. The commercial market also includes various activities of a non-industrial nature such as horticulture, bread baking, timber seasoning, pasteurisation, etc.

For convenience sake, the use of natural gas for transportation purposes is also described in this chapter as it is considered to be more closely aligned to a service type activity than to other market sectors.

Market Scope

The commercial market is a significant market for gas contributing typically about one-quarter of total gas utility sales in well established markets where space heating and/or air conditioning is an important use of gas.

Many of the commercial applications for gas are basically scaled up versions of those to be found in the residential market. While commercial appliances may be larger and more sophisticated than their residential counterparts, their purpose, i.e. for cooking, water heating, space heating, air conditioning, etc., is essentially the same. For example, gas is a popular fuel for the preparation and cooking of food in catering establishments because of its speed, flexibility, versatility, visible flame and cheapness compared with electricity, its main competitor, even though there are some applications, e.g. microwave ovens, where gas cannot compete.

Gas absorption air conditioning units are used extensively in commercial buildings in Japan, increasingly in the United States and to an extent elsewhere. They can provide efficient cooling without emitting harmful chlorofluorocarbons (CFCs), whereas many electric chillers currently in service can emit CFCs.

Gas is increasingly preferred over oil for space heating as many commercial users are located in densely built up urban areas where space for oil product storage tanks is difficult to find or can be better used for other purposes. Moreover, depending upon the oil product and combustion equipment used, emission levels may be a problem. Gas also offers the customer a much wider range of appliances.

Another example is co-generation in which gas is used to generate electricity with a gas engine or, for larger consumers, with a gas turbine with the excess heat being

Figure 12.1 **A Gas Turbine Co-generation System**

Courtesy of Tokyo Gas Co. Ltd.

used directly to warm air or to produce steam for heating purposes. Overall thermal efficiencies of about 80 per cent can now be achieved. An illustration of the growing popularity of co-generation systems with commercial customers is the seven-fold increase in the number of units installed over the last five years in Japan which now collectively generate nearly 450 MW.

Other recent developments such as heat pumps, radiant heaters, pulse combustion and fuel cells, all of which are described in other chapters, are particularly suitable for commercial uses.

Accordingly, to avoid unnecessary repetition, further discussion of the commercial market is confined to the following two examples.

Horticulture

Natural gas is used extensively in Europe by greengrowers for water heating, air heating and carbon dioxide production. In the Netherlands alone some 10500

Figure 12.2 **A Gas Engine Heat Pump System**

Courtesy of Tokyo Gas Co. Ltd.

growers consume over 2.5 Bcm p.a. of natural gas to heat their greenhouses, which cover an area of 8000 hectares.

The majority of commercial greenhouses are heated by hot water tubes which give uniform heat distribution throughout the greenhouse, important for even growth. By employing two heat exchangers and two heating circuits the efficiency of a hot water boiler can be increased from 80 to 95 per cent. The tubes of the low temperature circuit are located either in the ground for root heating or just above it for bed heating, while the high temperature tubes heat the space above the plants. Gas has the advantage over most other fuels of not producing any particulates which can deposit on the glass of greenhouses blocking out the light and corroding the metal frames.

Gas-fired air heaters offer similar advantages. However, as the return air temperature is never higher than the greenhouse air temperature only one heat exchanger is required. Hence combustion gases can be cooled to even lower temperatures giving efficiencies of up to 97 per cent.

Carbon dioxide can contribute to good plant growth. Normal ambient air contains 0.034 per cent by volume but this can drop by about one-half in an unventilated greenhouse. Many greengrowers consider the optimum CO_2 content to be 0.10 to 0.15 per cent and this can be achieved by trickle feeding compressed CO_2 supplied in cylinders but this is expensive. Alternatively, CO_2 can be drawn off by a fan from the flue of the greenhouse boiler and distributed throughout the greenhouse. The flue gas has to be continually monitored for the presence of carbon monoxide for the safety of the greenhouse workers. Also the presence of CO is an indication of incomplete combustion with the probability that ethylene has been produced in sufficient quantities to affect plant growth adversely. Low NO_x burners are required to reduce the damaging effect of these acidic emissions on the plants being grown. Air heaters specifically designed to produce CO_2 are used when the main heating boilers are not required.

Other ways being developed for greenhouse heating include heat pumps, combined heat and power systems, submerged combustion and radiant heaters.

Absorption Type Chiller/heaters

Double-effect absorption chiller/heaters provide air conditioning for offices, factories and other large buildings, or chilled and hot water for use in production processes. One such system is described below - see also Figures 12.3 and 12.4.

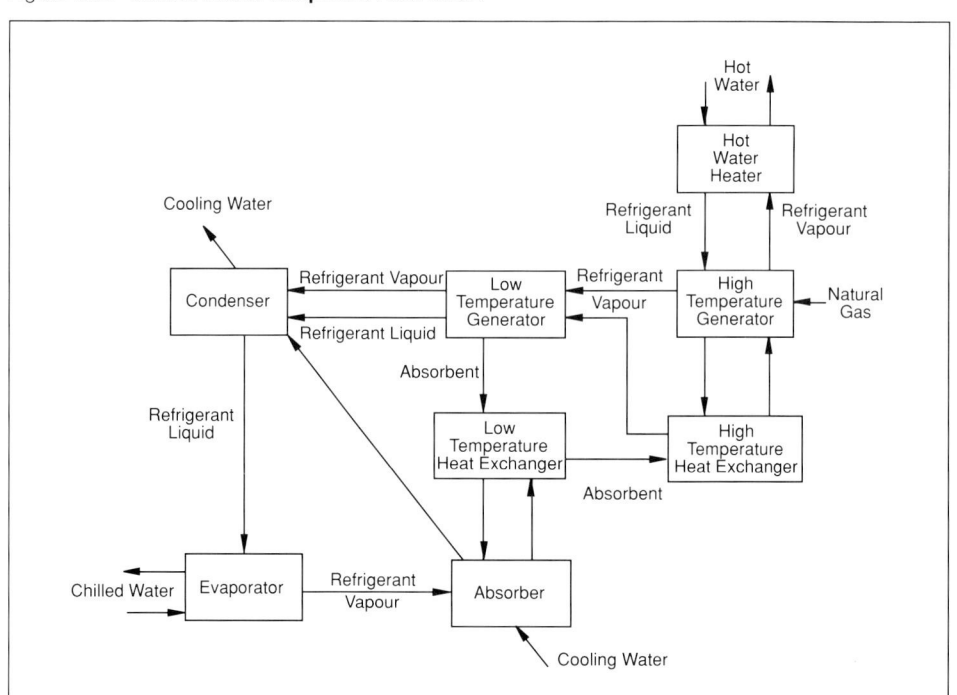

Figure 12.3 **Chiller/heater Simplified Flow Chart**

In the cooling cycle refrigerant dispersed in an evaporator extracts heat from the water to be chilled passed through it in tubes and is vaporised. In the absorber, concentrated absorbent is dispersed on tubes through which cooling water is passed. The concentrated absorbent is then diluted by the vaporised refrigerant and pumped to a high temperature generator via low and high temperature heat exchangers to be reconcentrated. In the high temperature generator the diluted absorbent is heated by natural gas to boil out the absorbent which bubbles up to a separator where it is separated from an intermediate absorbent. The vaporised refrigerant generated in the separator passes to a low temperature generator to give off heat to the intermediate absorbent before passing to the condenser, then to the evaporator to begin a new refrigerant cycle. The absorbent, which has been reconcentrated in the low temperature generator, passes through the low temperature heat exchanger on its way to the absorber to begin a new absorbent cycle.

Figure 12.4 **An Absorption Chiller/heater Unit**

Courtesy of Sanyo Electric Trading Co. Ltd.

In the heating cycle, diluted absorbent flowing into the high temperature generator is boiled by natural gas. The absorbent and vaporised refrigerant flow into the separator, then through a cooling/heating selection valve before being fed into the evaporator which functions as a condenser. Hot water in the evaporator tubes picks up heat given off during condensation of the refrigerant. The condensed refrigerant falls into a pan, overflows, falling to the bottom of the absorber and mixing with the concentrated absorbent. This mixture is pumped through low and high temperature heat exchangers back to the high temperature generator.

Advantages compared with the conventional centrifugal chiller and boiler systems include smaller space required as a separate boiler is not needed, fuel consumption is about 40 per cent less, electricity is only necessary to operate small circulating pumps and a combustion blower, can be operated at any load with efficient partial load performance, there are few moving parts, easier to maintain and low vibration and noise levels.

Automotive Uses

Manufactured gas was first used as an automotive fuel during World War I to help conserve conventional liquid fuels. Three main systems were developed. One was to store the gas in large inflatable bags located on top of the vehicle from where it was piped to the engine. Another system was to directly produce the gas on a detachable trailer towed by the vehicle. Subsequently, gas was compressed into cylinders. On-board storage pressures of up to 340 bar (5000 psi) were achieved. Despite their success, these systems were abandoned shortly after the war only to be revived, with improvements, during World War II.

The use of compressed natural gas (CNG) as an automotive fuel originated in Italy in the 1930s which is now the world's largest market for this application – see Table 12.1.

Table 12.1 **Natural Gas Vehicle (NGV) Statistics**

Country	Vehicles Converted	Refuelling Stations
Italy	235000	240
USSR	200000	300
Argentina	100000	125
New Zealand	50000	350
USA	30000	330
Canada	26000	170
Other countries (19)	1500	30
Totals	642500	1545

Virtually all these CNG vehicles have converted gasoline engines; it is estimated that their total annual consumption of natural gas is less than 1.5 billion m^3. To put these statistics into perspective, the United States has a vehicle population of some 195 million; if only one per cent of the gasoline vehicles in OECD countries were converted to natural gas, this would represent a total consumption of about 5 billion m^3 p.a.

Natural gas has a number of advantages over gasoline (and diesel) as an automotive fuel. It has a high octane number (over 120) which would permit compression ratios in the range of 12 to 15:1 and resultant high efficiency with a purpose-built engine. Because of its relatively low carbon content, natural gas is a cleaner burning fuel with the result that engine wear is low with practically no carbonisation of the spark plugs and internal components between services; engine oil contamination and dilution is also reduced. As methane has a high volatility, reliable and efficient mixing of air and fuel permits leaner mixtures and high thermal efficiency. Flame speeds in stoichiometric methane/air mixtures are slower than those with gasoline so that with optimised (advanced) spark timing fuel consumption is improved.

Polluting emissions are substantially reduced with natural gas. Sulphur dioxide emissions are virtually negligible, typically less than 1 mg/MJ compared with over 21 for gasoline and over 88 mg/MJ for diesel engines. As natural gas usually contains no nitrogen compounds, nitrogen oxides as products of combustion can be as low as 0.1 mg/MJ in contrast with diesel engines which are about 8 times greater. During stoichiometric combustion of the gas no hydrocarbons or ash are produced, while carbon monoxide emissions can be over 100 times lower than a gasoline engine. In essence, a gas-fuelled engine only produces three products of combustion in any quantity, nitrogen oxides, harmless water vapour and carbon dioxide. However, minor amounts of unburned methane may also be produced. Tests suggest that emissions of carbon dioxide are more than 30 per cent lower than for gasoline engines. Emissions of lead or benzene do not exist as these additives are not used nor are they necessary with gas-fuelled engines.

There are some other advantages but equally there are several disadvantages the most important being shorter distances before refilling is required, reduced engine power, the loss of carrying space in the vehicle to accommodate the CNG storage cylinders and the cost of conversion.

Almost all natural gas fuelled vehicles in current service are gasoline engined converted to use CNG as well as gasoline, or occasionally converted to use CNG only. However, increasing numbers of 'dedicated' vehicles designed and built for CNG use are now being produced notably by the three major American automobile manufacturers.

Although about 500 vehicles with diesel engines have been converted, as methane has a low cetane quality it is inherently unsuitable for use in a diesel engine without using a pilot quantity of diesel (typically 5 to 10 per cent of the full power requirement) to facilitate ignition.

Power losses in dual fuelled CNG vehicles can be in the range of 5 to 30 per cent. These losses are reduced when the vehicle is optimised to run on CNG. If acceleration times are not too important for the vehicle's normal duty, power losses are compensated by an improvement in energy efficiency of up to 15 per cent due mainly to the efficient combustion of the fuel/air mixture.

The extent to which the loss of space to house the CNG cylinder(s), in addition to the existing gasoline tank, is a disadvantage will depend upon the vehicle's duty.

Conversion costs vary with the type of vehicle involved, local labour and equipment costs, etc., but typical rates for the United States are around $1000 for the conversion kit, $400 to $500 for storage cylinders and up to $700 for labour giving a total of some $2500 to $3000. A refuelling station capable of handling up to 100 vehicles per day could cost around $250000.

Figure 12.5 **A CNG Refuelling Station In Italy**

As far as safety is concerned, CNG is considered to have several advantages over liquid fuels. Refuelling is through a sealed, fail-safe system so no vapours are released to atmosphere. CNG storage cylinders are considerably stronger and usually better sited on the vehicle than mild steel liquid fuel tanks. In the unlikely event of a CNG cylinder leaking after an accident, the gas would rise and diffuse rapidly whereas liquid fuels would spread over a wide area, with a high risk of ignition, before eventually evaporating to form a slowly dispersing heavy vapour. CNG fuel lines are made of high tensile steel which is very difficult to sever. Methane has a wider flammability range (5 to 15 per cent volume) in air than gasoline (0.5 to 8 per cent) but is less likely to ignite at very weak mixtures. The temperature at which methane ignites spontaneously in air and continues to burn is over $650°C$ compared to $220°C$ for gasoline. CNG has enjoyed a first class safety record to date.

Figure 12.6 **Overnight CNG Trickle Feed Refuelling**

The economics of CNG are crucially dependent on actual conversion costs, the differential between the into-tank prices of natural gas and gasoline (or diesel), average annual mileages and government attitudes to the substantial tax usually levied on automotive fuels. One of the reasons for the success of CNG in Italy is its cheapness, about half the price of diesel and a quarter that of gasoline on an equivalent energy basis. However, cost considerations may not be the only criterion, for example, in situations where reducing vehicle emissions is a prime objective for environmental/health reasons. In some countries, government monetary incentives are available to partly offset the cost of conversion and/or for the building of refuelling stations.

For operating cost and/or environmental reasons CNG conversion can be an attractive option for urban bus and truck fleets, taxis, company car fleets and, as and when a countrywide network of refuelling stations is established, for long distance transport. Private car owners may also find CNG conversion attractive if they undertake relatively high mileages. Expansion of this market sector will be helped by the development of overnight trickle feed household refuelling appliances which have been successfully and safely tested in several countries – see Figure 12.6.

In addition to CNG, various small scale trials have taken place using LNG as an automotive fuel. Its advantages and drawbacks as a fuel for this application are, of course, essentially the same as for CNG but from the limited data and experience available conversion kits would be more expensive. LNG's main drawback is that unlike gaseous natural gas, it is only available at a few locations in a limited number of countries. Whereas the provision of compressor capacity is straightforward and relatively inexpensive, providing liquefaction capacity at conveniently accessible locations would be much more difficult and expensive. But an important compensation viz-a-viz CNG is the much greater energy intensity of LNG which would give a far greater vehicle range for a given tank size.

Undoubtedly, the use of CNG will grow and spread geographically but even under favourable circumstances it will tend to complement rather than replace conventional liquid fuels for many years to come.

While LNG for automotive purposes may have a limited potential in site specific locations for purpose-built vehicle fleets, it seems unlikely that it could compete with CNG on any scale.

One area of current interest for LNG as a transport fuel is for railway locomotives and heavy duty 'off-road' transport such as large vehicles used in mining operations. Here dedicated route patterns and constant yet heavy fuel use enables an LNG liquefier to be operated at a high load factor with attractive economics under appropriate conditions. Trials to date in the United States with several diesel-engined locomotives converted to LNG have been successful.

Chapter 13

THE INDUSTRIAL MARKET

The number and variety of energy uses in the industrial market are almost limitless. They range from relatively straightforward under-boiler steam raising to applications requiring very precise heating for the manufacture or treatment of high quality end-products. The variety of fuels that can be used – natural gas, fuel oil, gas oil, LPG, kerosene, electricity, coal, etc – for many applications is also quite large.

To put it very simply, the industrial market is essentially all about cost effectiveness and improving the efficiency of heat transfer for the application in question. More recently, environmental considerations in many countries have become equally important. In the case of fossil fuels, governments are setting targets to be realised by specified years for reductions in emissions of sulphur dioxide, particulates and nitrogen oxides (NO_x). While all fossil fuels produce NO_x when burnt, and the higher the temperature the greater volume of NO_x that is produced, natural gas produces less NO_x than most other fossil fuels, negligible sulphur dioxide and no particulates. Nevertheless, in order to meet future emission standards and to ensure natural gas maintains its competitive edge, considerable research and development work is focused on improving heat transfer efficiency at lower levels of emission.

As it is impossible to cover the whole range of industrial applications where natural gas can be used, several well-established uses and new developments have been selected to illustrate how gas is tackling the challenges indicated above.

Radiant Burners

Radiant burners work on the principle of forcing a gas-air mixture through the porous surface of the burner and igniting it on the outside. This results in a carpet of minuscule flames which burn close to and within the surface so that a significant proportion of the heat is conducted into the material of the burner. This heat is then radiated away to the surroundings, effectively cooling the flames and providing valuable radiant heat in addition to the convected heat output. The merit of this process is that low flame temperatures produce much lower levels of NO_x, typically 80 per cent less than for a similar size gas-fired conventional burner.

Apart from emission considerations, radiant (infra-red) heating is particularly important for certain industries, e.g. for the making and processing – drying, coating and moulding – of paper. Indirect steam heated drying is the conventional (and inefficient) paper drying technology which has been increasingly replaced by

Figure 13.1 **Walking Beam Type Reheat Furnace**

radiant gas and electric driers particularly for rapid drying. Recently, gas-fired, surface-combustion, radiant burners based on ceramic fibres and particularly metallic fibres have been developed and used successfully, offering an attractive alternative to electric heaters. The remaining discussion in this section is confined to metallic fibre burners (MFB).

The combustion surface of MFBs comprises fine metallic fibres sintered into suitable high strength mats of high porosity. Mats of this type have been developed by Shell Research in co-operation with Bekaert of Belgium and are now incorporated into a variety of burners produced by a number of manufacturers. The advantages of MFBs include:

- *very rapid (a matter of seconds) heat-up and cooldown which helps to minimise fire hazards in, for example, a paper mill;*
- *uniform surface heating over a wide range of thermal inputs;*
- *high turndown capability without flashback;*
- *very tolerant of variations in gas compositions;*
- *fast response time when the gas/air mixture is changed;*
- *robust and durable in both the blue flame and radiant modes of operation;*
- *easy to clean and maintain.*

An important advantage of MFBs is the very low levels of NO_x they produce not only in the radiant mode, but also in the high intensity blue flame mode. While NO_x emissions start to increase once patches of blue flame start to appear on the burner surface, these can be reduced with appropriate amounts of excess air. The point at which this transition from radiant to blue flame commences is known as the 'blue flame limit'. However, in many applications only convective heating is required for which blue flame operation is quite acceptable. Perforated burner designs facilitate low NO_x performance with high intensity blue flame operation.

With metallic fibre mats of about 4 mm thickness, open-air surface temperatures of up to $1150°C$ are possible in radiant mode operation of 650 kW/m^2. Higher intensities of up to 3000 kW/m^2 can be achieved in the blue flame mode.

Radiant burners are now available or are being developed for a variety of applications including space heating in industrial buildings and outdoor areas, industrial steam boilers, drying processes in the food and textile industries, heating and CO_2 enrichment in commercial greenhouses and residential central heating boilers.

Direct Reduced Iron

Direct reduced iron (DRI) at some 20 million tonnes per year is becoming a significant feedstock for steel production. Most DRI production is consumed locally in countries such as Venezuela, Mexico, the Soviet Union, Argentina, Indonesia, Saudi Arabia, etc. Over 90 per cent of world DRI production comes from natural gas-fired plants consuming 5 to 6 Bcm a year in total. Because of its purity and absence of other metals, DRI is very suitable for the production of high quality steel. Growth prospects for DRI are considered to be very good.

Direct reduction processes, which essentially involve carbon monoxide and hydrogen produced from natural gas, reduce iron ore by removing the associated oxygen at temperatures below the melting point of any materials in the process – i.e. below about $1000°C$. Natural gas processes are based on the reduction of iron oxide, usually hematite (Fe_2O_3), by synthesis gas rich in carbon monoxide and hydrogen – see Chapter 15 re synthesis gas production.

Figure 13.2 **Direct Reduced Iron : Midrex Process Flow Diagram**

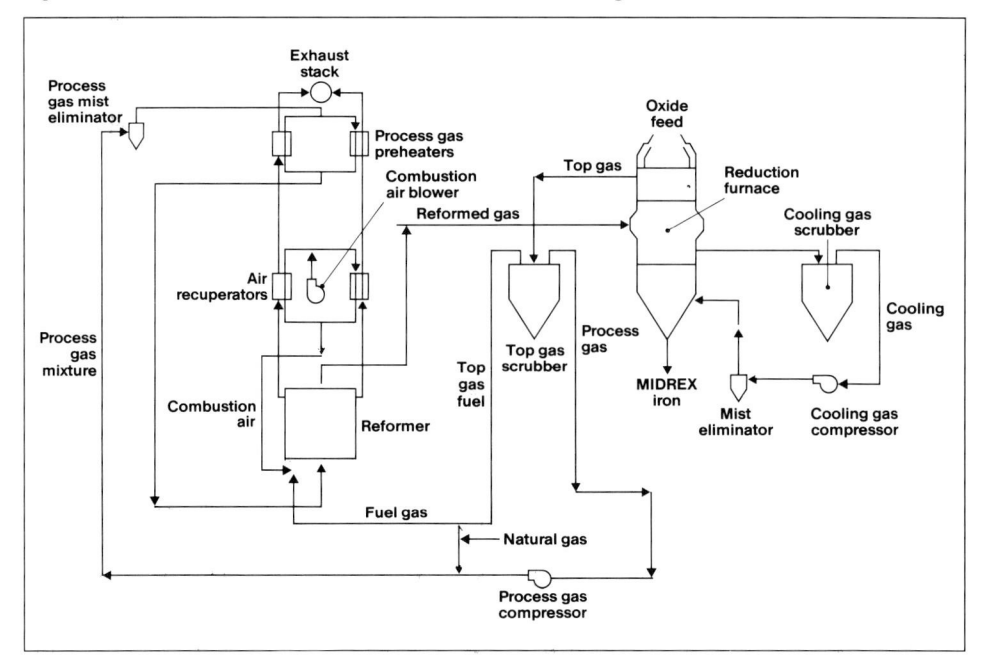

In a highly reduced product, iron is in the form of metallic iron or the partially reduced form of iron oxide, wustite (FeO). Typical reduction reactions are:

To metallic iron $\quad Fe_2O_3 + 3CO \longrightarrow 2Fe + 3CO_2$

and

$$Fe_2O_3 + 3H_2 \longrightarrow 2Fe + 3H_2O$$

To wustite $\quad Fe_2O_3 + CO \longrightarrow 2FeO + CO_2$

and

$$Fe_2O_3 + H_2 \longrightarrow 2FeO + H_2O$$

There are a number of proprietary processes of which the Midrex process currently has over 60 per cent of the market. The main components of the Midrex process are the direct reduction shaft furnace, gas reformer and cooling-gas system – see Figure 13.2.

Reducing process gas (H_2 and CO) enters the reducing furnace through a bustle pipe and ports located at the bottom of the reduction zone and flows up through the descending solids. Iron oxide reduction takes place at about $900°C$. The largest portion of the top gas is recompressed, enriched with natural gas, preheated ($400°C$), and piped into the reformer tubes. In the catalyst tubes, the gas mixture is reformed to CO and H_2 which is then recycled to the reducing furnace at over $900°C$.

The excess top gas provides fuel for the burners in the reformer. Hot flue gas from the reformer is used in the heat recuperators to preheat combustion air for the reformer burners and also to preheat the process gas before reforming. The addition of heat recuperators to these gas streams has enhanced process efficiency which is now around 2.5 million kcal/tonne of DRI.

Cooling gases flow through the burden in the cooling zone of the shaft furnace. The gas then leaves at the top of the cooling zone and flows through the cooling-gas scrubber. The cleaned and cooled gas is compressed, passed through a demister and recycled to the cooling zone.

Oxygen Enrichment

Oxygen for combustion systems is usually obtained by supplying ambient air to the burners. Oxygen enrichment can be achieved in three main ways by:

- *adding oxygen to the combustion air up to an overall oxygen content of about 30 per cent before it enters the burner;*
- *injecting oxygen directly into the flame by means of a lance inserted through the burner or below it; and,*
- *replacing the conventional burner with an oxy-fuel burner.*

The main advantages of oxygen enrichment are:

- *it reduces the quantity of gas required for the same amount of end-product and hence user's fuel costs;*
- *it reduces exhaust gas emissions (notably CO_2) and sensible heat losses;*
- *it increases flame temperatures and radiant and convection heat transfers; and,*
- *it improves flame stability.*

Disadvantages are increased noise and NO_x emissions and, of course, the cost of supplying oxygen. However, the first two can be largely mitigated by enclosing the burner, improved burner design and flue gas recirculation.

In summary, oxygen enrichment is attractive to existing and prospective gas consumers by offering fuel savings of up to some 50 per cent and reduced heating or furnace residence times.

Regenerative Burners

These compact burners operate in pairs and can recover up to 90 per cent of the heat in the exhaust gases. Each burner has its own heat exchanger filled with ceramic material. They are fired alternatively every few minutes and hot gases pass into the working chamber from one burner and leave via the other. Heat retained in the ceramic material pre-heats the combustion air up to 1200°C.

Figure 13.3 shows a 12 tonne capacity holding and melting furnace for melting

Figure 13.3 **Scrap Aluminium Melting**

Courtesy of British Gas PLC

scrap aluminium which was originally oil-fired. After conversion to gas firing using a twin regenerative burner system, energy consumption dropped by 45 per cent, productivity increased by 16 per cent, emissions were reduced and maintenance decreased by 100 man-hours per week.

Fuel Cells

Existing methods of generating electricity on a commercial scale are relatively complex, noisy and, if generated from fossil fuels, produce various pollutants. Moreover, even with state of the art technology, they are not particularly efficient.

In essence, fuel cells convert the chemical energy of a fuel, i.e. natural gas in this case, into electricity by passing hydrogen continuously over a suitable porous plate (an electrode) separated by an electrolyte (an electrically conductive medium) from a second plate over which oxygen is passed. This creates a small voltage across the plates. If a series of such cells are stacked together, a useful direct current power output is created. The direct current generated can then be converted into alternating current by passing it through a converter.

Various electrolytes can be used, which operate at different temperatures. The most important are alkaline ($80°C$), phosphoric acid (150-$200°C$), molten carbonate ($650°C$) and solid oxide ($1000°C$). Although alkaline cells are reported to have

Figure 13.4 **Molten Carbonate Fuel Cell Power Plant, San Ramon, California**

Courtesy of Pacific Gas & Electric

almost reached the commercial stage, they have the disadvantage of requiring very pure hydrogen and oxygen, whereas phosphoric acid and molten carbonate cells can operate on a hydrogen-rich gas. This can be obtained by passing natural gas and steam over a suitable catalyst. The heat for this process can be provided by burning the waste gas from the fuel cell, which contains residual amounts of hydrogen, methane and carbon monoxide, supplemented with an injection of natural gas. Steam can be produced by using some of the heat given off by the cell itself, while the remaining heat can be used for other purposes. Thus fuel cells offer attractive prospects for combined heat and power applications.

Other advantages over existing fossil fuel electricity generation are higher efficiency (up to 80 per cent) whether on full or part loads, low flame temperatures with consequential lower NO_x emissions, noise-free and unattended operation, very little water is used, they take up little space, and they can be easily expanded by simply adding more stacks. Furthermore, the reaction in the cell itself produces only water.

The main disadvantage is cost. With present technology fuel cells are substantially more expensive per kilowatt output than conventional means of generating electricity from fossil fuels. However, simpler system designs and lower costs are considered to be just a matter of time. A number of companies expect to have commercial units of 50kW to 2 MW capacity in service from the mid 1990s. In 1991,

Pacific Gas & Electric brought into operation the world's first molten carbonate type fuel cell producing initially 20 kW to be phased up to 100 kW – see Figure 13.4. Demonstration plants of the phosphoric acid type with outputs of 25-50 kW are also being tested in Europe and Japan, and designs for larger units are well advanced.

Lime Kilns

Coke has been the traditional fuel for firing simple vertical shaft lime kilns, but as suitable grades of coke become scarcer and more expensive opportunities exist to replace coke with natural gas, especially as conversion costs are modest.

Quicklime or calcium oxide (CaO) is produced by burning limestone or calcium carbonate ($CaCO_3$) in a kiln. The process is essentially endothermic. The reaction is:

$$CaCO_3 \longrightarrow CO_2 + CaO$$

Limestone is first heated to the reaction temperature which is at least $900°C$ and then maintained at that temperature until the reaction, i.e. calcination, is complete. However, as pure calcium carbonate is seldom available, not all the limestone is converted during calcination to quicklime with the result that some limestone remains in the quicklime produced. Control of the burning process helps to determine the quality of product obtained.

Heat requirements depend largely on the specific heat of the limestone and hence its $CaCO_3$ content. Part of the heat supplied during the burning process is recovered within the kiln, part is lost to atmosphere with the flue gas emissions, and part is lost through the kiln walls, resulting in overall efficiencies of about 70 to 85 per cent. Typical thermal requirements average out at around 1000 kcal/kg (4.2 MJ/kg); for complete conversion of pure $CaCO_3$ the heat requirement would be about 700 kcal/kg.

As the thermal requirements of using coke or gas are much the same, the economics of using either fuel will largely depend on the going prices of the two fuels. However, apart from economic considerations, replacing coke by natural gas offers a number of other advantages. As gas is ashless and sulphur free it enables a higher quality quicklime to be produced. Flue gases from natural gas burning are more environmentally acceptable – no SO_x, less NO_x and no fly ash. Unlike coke, no storage facilities and handling equipment are necessary. Gas is able to respond much quicker to variations in processing conditions and it facilitates partial and/or full automation of burner and kiln operation.

The Emission Challenge

In many countries limits on allowable pollutant emissions arising from industrial processes are becoming ever more stringent. Flue gases produced when natural gas is combusted contain virtually no SO_x, particulates, uncombusted hydrocarbons or fluorides. Emissions of NO_x are also lower from natural gas than from other fossil fuels, nevertheless considerable effort is being devoted to reducing them still further.

NO_x is essentially the product of the reaction of atmospheric oxygen and atmospheric nitrogen at the temperature of combustion and is not fuel-induced, especially as most natural gases contain virtually no nitrogen. The formation of NO_x is mainly a function of flame temperature, the partial pressure of oxygen and the residence time of the products of combustion in the combustion chamber.

Important aspects for the control of NO_x are the design of burners and how they are sited in the combustion chamber, the design of the combustion chamber, the fuel-air ratio, combustion air temperature and the load in the combustion chamber. Some examples of how NO_x emissions can vary or be varied are:

- *tangential firing produces much less NO_x than horizontal firing;*
- *recuperative furnaces produce less NO_x than regenerative furnaces as the combustion air is preheated to a lower temperature;*
- *the short flames of a non-aerated burner are usually hotter than the long flames of a pre-mixed burner and produce more NO_x;*
- *as combustion temperatures rise NO_x emissions increase exponentially;*
- *NO_x emissions increase with fuel-rich mixtures as the proportion of air in the mixture approaches the stoichiometric ratio, NO_x emissions then fall markedly as the excess air level is increased;*
- *the formation of NO_x usually declines as the load in the combustion chamber decreases;*
- *as heat transfer by radiation is lower in the case of a gas flame than for an oil flame, the temperature of a gas flame is higher and more thermal NO_x is formed.*

The efforts of burner and combustion chamber manufacturers to reduce emissions by NO_x formation control and by post-formation clean-up tend to be focused on:

- *designing low NO_x burners;*
- *reducing the preheated combustion air temperature, although this is partly offset by increased fuel consumption;*
- *increasing the amount of excess air;*
- *staging the injection of fuel and combustion air in large combustion chambers;*
- *developing flue gas recirculation systems; and,*
- *optimising the design of the combustion chamber and the geometry of the burners in the chamber.*

The radiant burner described earlier is one important example of the progress being achieved in reducing NO_x emissions. Higher radiant output from the burner decreases flame temperature markedly thus reducing NO_x output.

Chapter 14

POWER GENERATION

Electricity generation is the largest single user of primary energy, accounting for over one-third of the world's total energy consumption. Demand for electricity has been rising over the last decade or so at a rate almost double that of energy as a whole, a trend that is expected to continue for the foreseeable future.

Thermal power plants - as distinct from nuclear, hydro and other means of generating electricity - consuming coal, oil products and natural gas, supply about 65 per cent of world electricity demand. Coal, with some 38 per cent of the market, is by far the major source of energy used for power generation with oil and gas contributing about 11 and 16 per cent respectively – see Figure 14.1. Although percentage contributions vary widely from one country to another, coal is still the dominant fuel for power generation in most industrialised countries even in those cases where ample supplies of natural gas have been available for many years. The Netherlands, where gas is the major fuel for power generation, is one exception to this generality.

This situation is now changing driven by such factors as the increasing concern for the environment, the development of new technologies and the growing availability of natural gas, all of which have caused governments to revise their energy policies. For example, in the United States the Fuel Use Act, which had effectively banned the use of gas in new power plants, was repealed in 1987. More recently in Europe, the EC Directive which discouraged the use of gas in power plants larger than 10 MW has been relaxed, and with the Large Combustion Plant Directive setting SO_x, NO_x and particulates emission targets for each EC country, gas for power generation is now very much in favour.

Today, about 380 Bcm, or some 20 per cent of total world gas consumption, is used for power generation. The expectation is that in both absolute and percentage terms this market sector will grow strongly in the coming years and probably outpace increases in other uses of gas.

As far as gas-related technology is concerned – environmental issues are discussed later – the key factor has been the development of large and reliable gas turbines. These are now used for power generation in a number of ways. New large scale combined cycle plants are being built at an increasing rate, while many existing power plants are being 'repowered' to increase efficiency and reduce NO_x and SO_x emissions. Some large scale systems also provide process steam to a willing user, so-

Figure 14.1 **World Power Generation Fuel Consumption 1990**

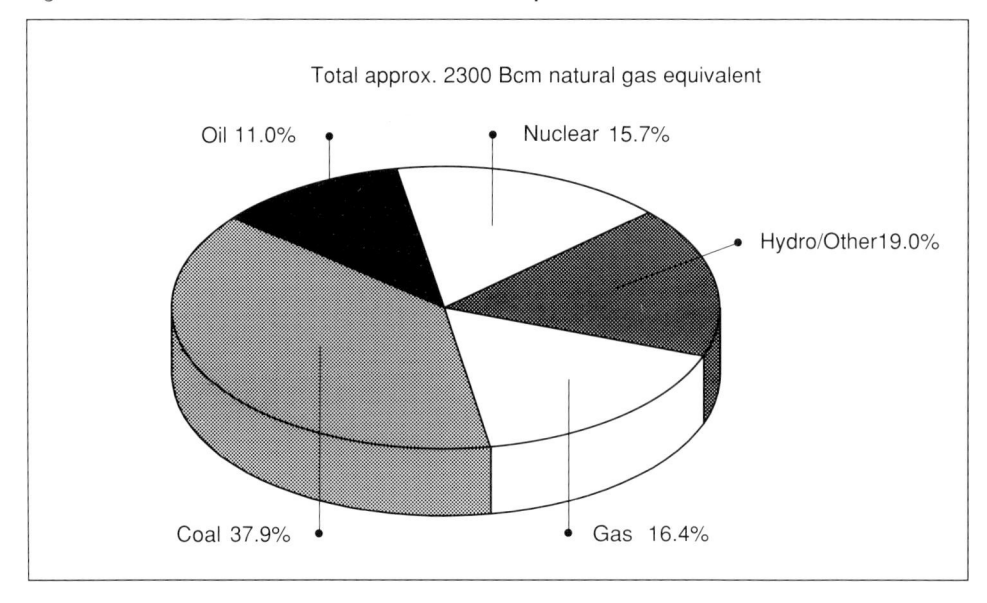

Total approx. 2300 Bcm natural gas equivalent

- Oil 11.0%
- Nuclear 15.7%
- Hydro/Other 19.0%
- Gas 16.4%
- Coal 37.9%

called 'combined heat and power (CHP)', although most use of CHP is with smaller aero-derivative gas turbines suitable for industrial use. Steam injected aero-derivative gas turbines are also used for industrial duties; steam injection reduces NO_x levels and increases gas turbine power output. However, the main focus of this chapter is on the biggest potential user of gas – combined cycle power generation.

Combined Cycle Gas Turbine

A gas-fired combined cycle plant is the most efficient of all the fossil fuel based technologies for generating electricity. In a combined cycle gas turbine plant – frequently referred to simply as CCGT – the heat contained in the exhaust gases from a gas turbine is passed through a heat recovery steam generator. The steam produced is then passed through a conventional steam turbine. Both the gas and steam turbines are connected to generators for the production of electricity – see Figure 14.2.

The thermodynamic advantage of a combined cycle plant lies in the large useful temperature drop between the heat input in the gas turbine and the exhaust heat in the condenser of the steam turbine. In a combined cycle plant this temperature difference is some 60 to 100 per cent higher than in a conventional steam turbine plant giving an overall increase in efficiency of up to 10 per cent.

In the past, the combined cycle has been limited in its application because of problems associated with the gas turbine – it was only available in relatively small sizes which would result in a multiplicity of these units being required to give a total power station capacity of say 500 MW. The reliability of the gas turbine when used

Figure 14.2 **Gas Combined Cycle Electricity Generation**

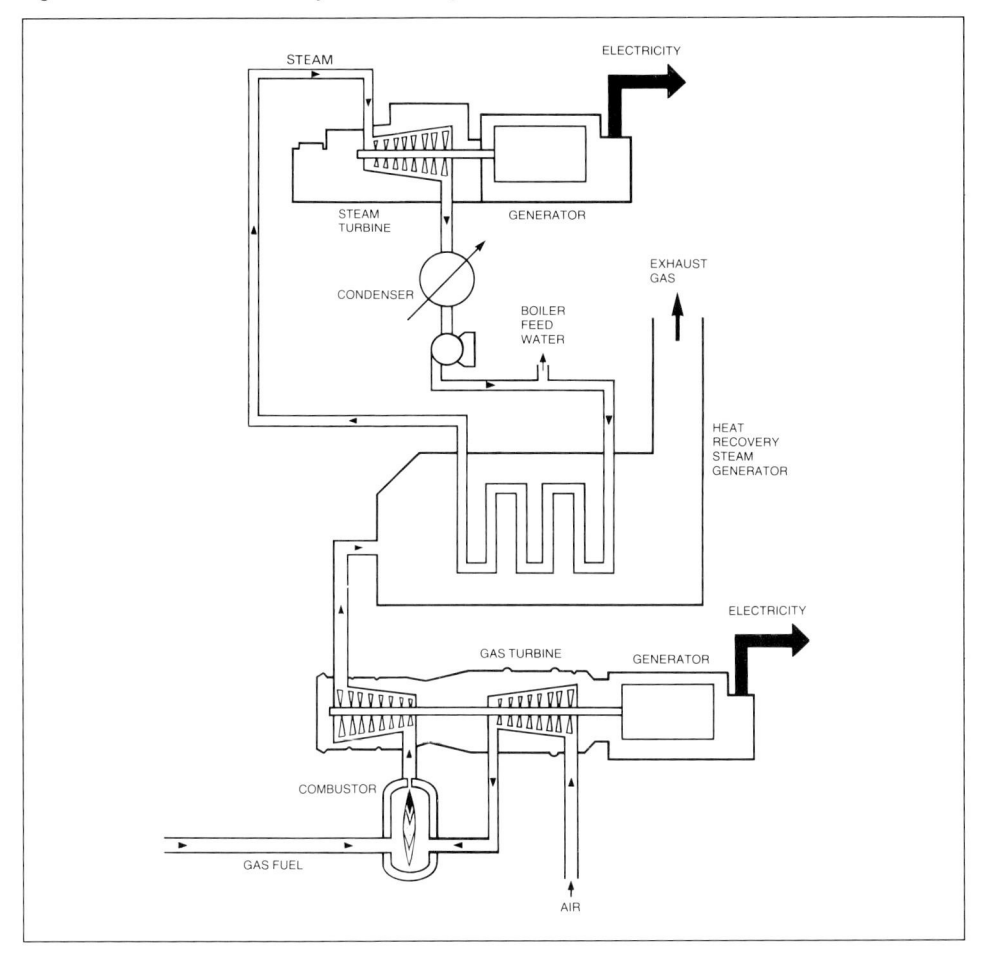

for peak power purposes was also inferior to that of the conventional steam boiler/steam turbine combination. However, recent technological advances now make it possible to construct gas turbines of up to 200 MW output and even larger units are feasible.

The most recently installed combined cycle plants are operating with availabilities of over 90 per cent. In addition, new construction materials and technologies enable gas turbines to operate at higher inlet temperatures, resulting in overall thermal efficiencies for the combined cycle which are currently about 52 per cent and expected to reach around 55 per cent within a few years. These efficiencies can be compared with the highest available efficiency from a conventional steam turbine power plant of 38 to 41 per cent, depending on the fuel used.

Figures 14.3, 14.4, 14.5 and 14.6 show the efficiencies of various components, the overall efficiencies of conventional coal, fuel oil and natural gas plants and that of a combined cycle plant. These percentages are typical of those which can be achieved with current technology and are after deducting allowances for electricity used within the plant. A modern-day combined cycle module usually consists of a number of gas turbines with one steam turbine. Optimum efficiency is achieved when the power output from the gas turbines is about twice that of the steam turbine.

These efficiencies compare the energy input of the fuel on a net calorific value (NCV) basis with the quantity of electricity sent out to the grid. They assume NCVs for coal of 26.0 MJ/kg, fuel oil 40.5 MJ/kg and gas 48.2 MJ/kg which is equivalent to 34.8 MJ/kg/m^3.

CCGT Cost and Other Aspects

In addition to environmental considerations which are discussed later, combined cycle plants have other advantages. These include shorter construction time, which

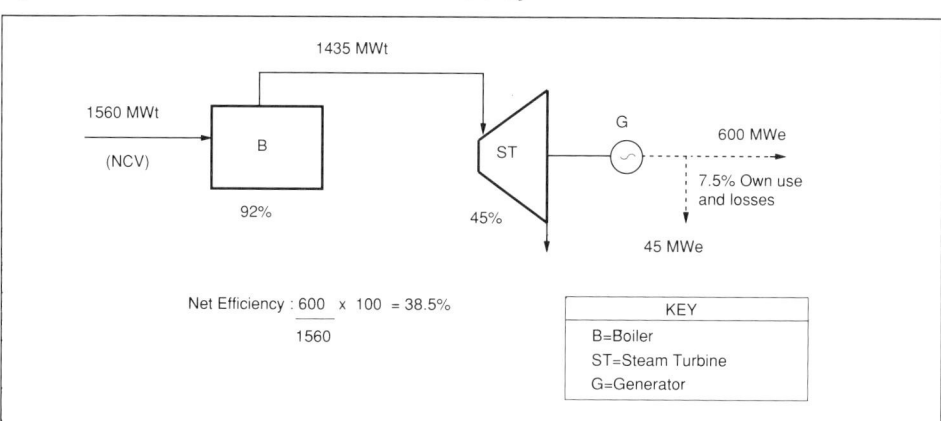

Figure 14.3 **Conventional Coal-Fired Plant Efficiency**

Figure 14.4 **Conventional Fuel Oil-Fired Plant Efficiency**

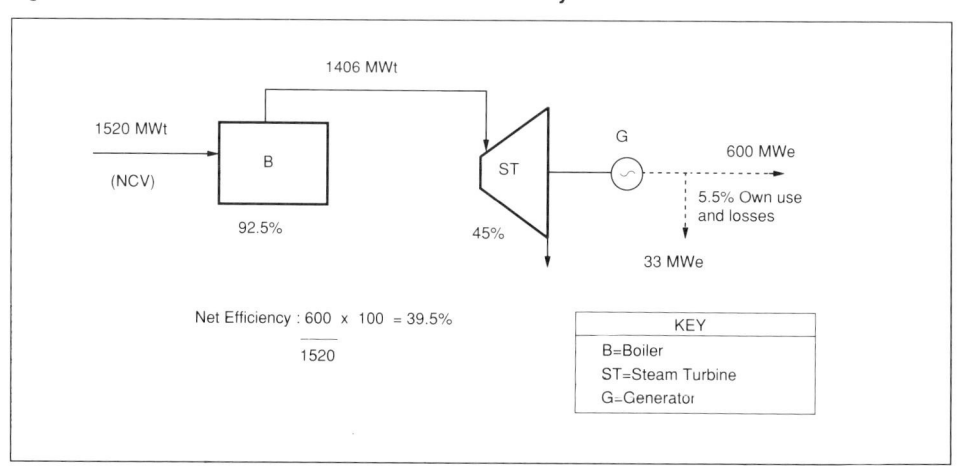

Figure 14.5 **Conventional Natural Gas-Fired Plant Efficiency**

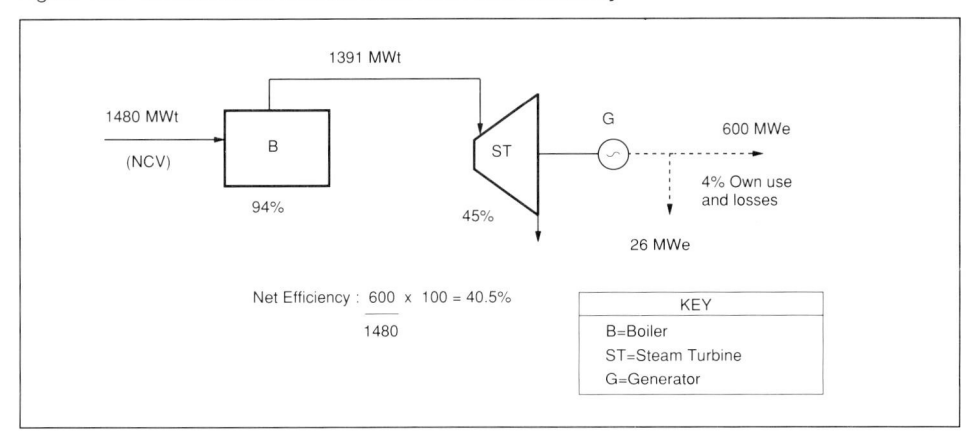

```
                         1391 MWt
   1480 MWt        ┌──────────────┐
                   │              │         G      600 MWe
   ──────▶    B    │      ST      │       ─○─ ─ ─ ─ ─ ─ ─▶
   (NCV)           │              │                4% Own use
                   └──────────────┘                and losses
         94%                45%
                                              26 MWe

         Net Efficiency :  600 × 100  = 40.5%         KEY
                           ─────────                  B=Boiler
                              1480                    ST=Steam Turbine
                                                      G=Generator
```

experience indicates should be at least one year less than for similar sized conventional plants under the same conditions, and much smaller plot sizes for the same output capacity. Capital costs in terms of dollars per kilowatt of output are normally lower for combined cycle plants for similar sites in the same country. Example figures of capital costs are not very meaningful as they soon become outdated. Moreover, cost comparisons are complicated by the need or otherwise to provide flue gas desulphurisation (FGD) facilities for conventional plants. This will depend on the quality of the coal or fuel oil used, local allowable emission levels and other factors.

Similarly, operating, maintenance and variable costs, including water provision and treatment costs, limestone (for FGD), gypsum and ash disposal costs, also vary

Figure 14.6 **Natural Gas Combined Cycle Plant Efficiency**

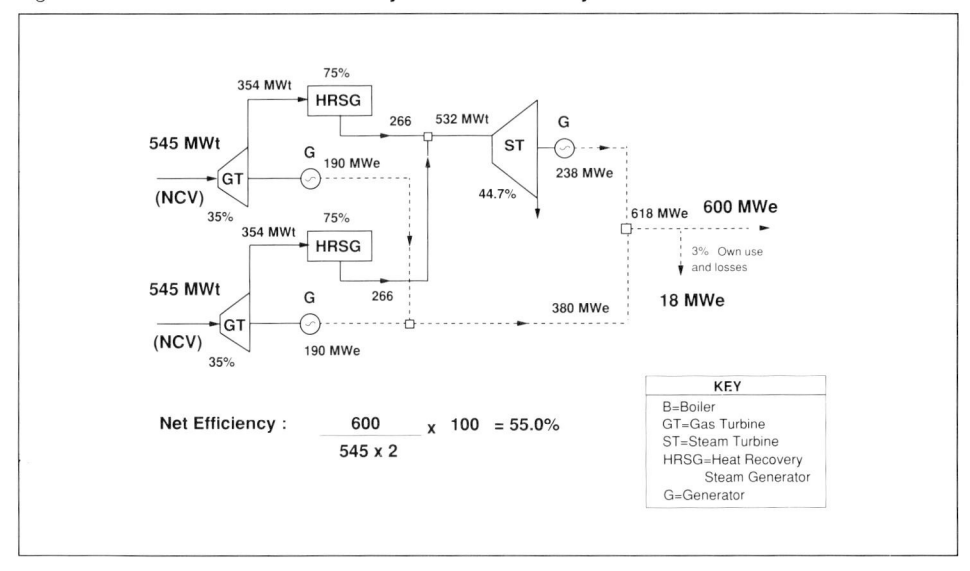

```
                              75%
              354 MWt     ┌────────┐
                          │  HRSG  │
                          └────────┘
                              │    266    532 MWt     G
  545 MWt                     ▼         ┌────────┐          238 MWe
 ─────▶  ┌──┐   G   190 MWe              │   ST   │   ─○─ ─ ─ ─ ─▶
  (NCV)  │GT├──○──────────────           └────────┘
         └──┘                              44.7%
    35%           354 MWt    75%                         618 MWe    600 MWe
                          ┌────────┐                   ┌ ─ ─ ─ ─ ─ ─ ─▶
                          │  HRSG  │                         3% Own use
                          └────────┘                         and losses
  545 MWt                     │    266
 ─────▶  ┌──┐   G             ▼            380 MWe       18 MWe
  (NCV)  │GT├──○──────────────
         └──┘      190 MWe
    35%

    Net Efficiency :      600      × 100  = 55.0%
                        ─────────
                         545 × 2
                                                    KEY
                                                    B=Boiler
                                                    GT=Gas Turbine
                                                    ST=Steam Turbine
                                                    HRSG=Heat Recovery
                                                           Steam Generator
                                                    G=Generator
```

substantially. Other variables are insurance and tax rates, plant utilisation rates (defined as the annual output of electricity expressed as a percentage of the maximum theoretical output at full load), plant input fuel costs, plant life, decommissioning and demolition costs at end of plant life, and the rate(s) of return sought by the electric utility for new investments.

In sum, the assessment of the capital costs and economics of combined cycle versus the alternatives, including the upgrading of existing depreciated plants, can only be evaluated on an individual, location-specific basis. All one can say is that evidence to date from many countries indicates that for broadly similar operating conditions, combined cycle plants have significant advantages over competitive systems before taking account of any over-riding environmental considerations that may apply.

Repowering

Existing conventional oil, gas or coal-fired power stations can be retrofitted or converted to take advantage of combined cycle features by the addition of one or more gas turbines. The exact design details will depend upon the original power station configuration and specification, but simplistically the gas turbine exhaust is routed to the original boiler, obviating the need for an air preheater for use as combustion air – see Figure 14.7.

The advantages of such a system include:

- *an increase in the overall plant efficiency from about 40 to some 46 per cent;*
- *a near constant part-load efficiency of 46 per cent when operating between half and full-load. This allows these plants to be operated efficiently as two shift, mid-range duty plants; and,*
- *a significant reduction in NO_x emissions from the power plant.*

In the majority of cases the existing boilers are retained, but it may be economic to replace them with specialised units to achieve maximum efficiency in the combined cycle mode.

Gas reburn can also be used in existing oil and coal-fired power stations. Reburning involves the injection of natural gas into the upper region of the boiler to create a rich zone where NO_x is converted to molecular nitrogen. Secondary air is then added to burn out any unreacted fuel at temperatures low enough to prevent further NO_x formation. Using 15 to 20 per cent natural gas in this reburning mode can reduce NO_x emissions by 50 to 60 per cent.

Steam or water injection is used frequently to reduce NO_x emissions from conventional power plants. With the smaller aero-derivative gas turbines used in industrial applications, it is possible to inject considerable quantities of steam, in fact all the steam produced in the heat recovery steam generator. In this mode, the gas turbine acts effectively as both a steam turbine and a gas turbine and produces a considerable amount of extra power. For example, a General Electric LM 5000

Figure 14.7 **Schematic Gas-Fired Repowered Combined Cycle Power Plant: One Unit**

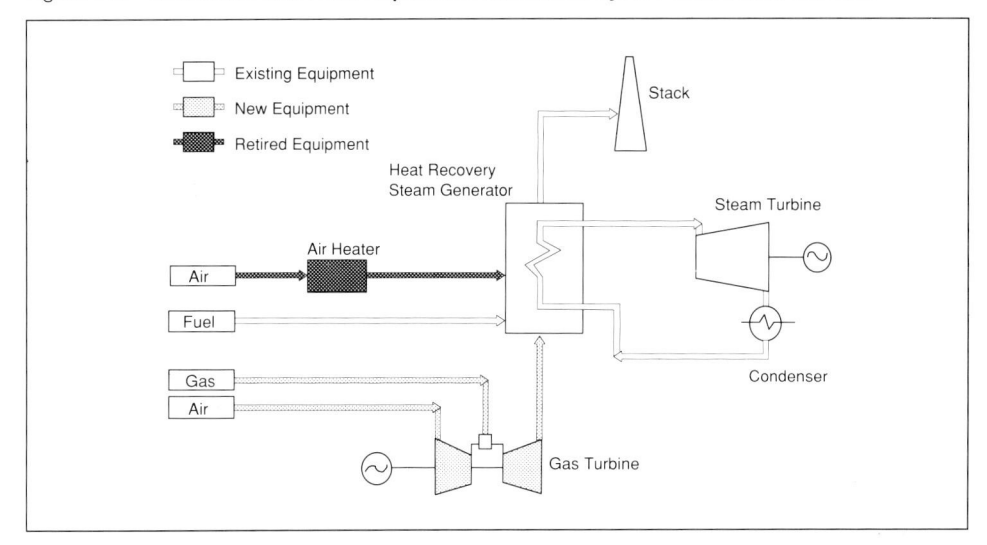

turbine will normally generate about 34 MW but when fully steam-injected this increases to 52.5 MW at 40 per cent efficiency.

Combined Heat and Power

Gas-fired combined heat and power (CHP) systems, also known as co-generation, are becoming very popular and span a wide spectrum of the market from relatively small scale residential and commercial consumers to large industrial plants. In essence, a CHP system produces steam or hot water for heating and process purposes and electricity for power.

For a large industrial plant, the combination of a gas turbine with a simple waste heat boiler that can retrieve much of the available heat to produce steam for process

Figure 14.8 **Typical Cogeneration Scheme**

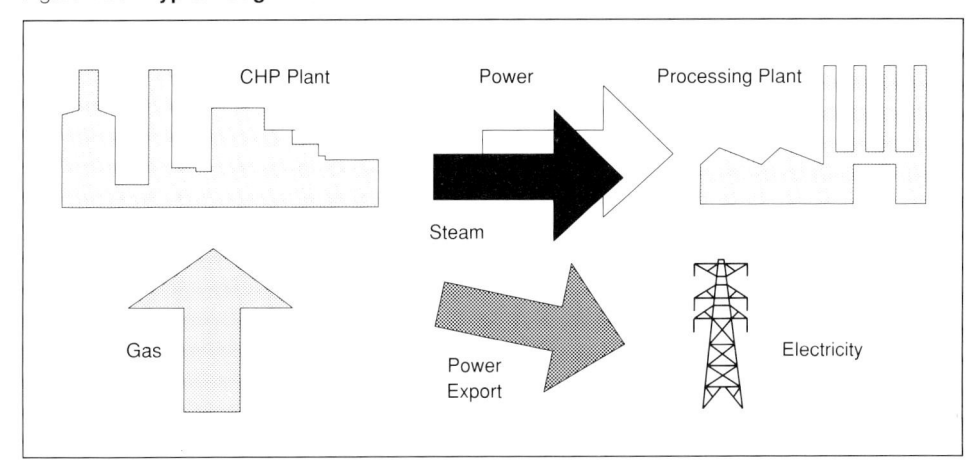

Figure 14.9 **Cogeneration Plant: Tokyo**

Courtesy of Diamond Gas Operation Co. Ltd.

needs, can result in an overall thermal efficiency of more than 80 per cent. Electricity generated that is surplus to the plant's own needs is usually sold to the local electricity utility thus producing a revenue to offset capital and fuel costs – see Figures 14.8 and 14.9.

At the other end of the spectrum, e.g. houses, apartments, hotels, offices, etc, the CHP concept is the same except that a gas turbine would not be practicable for small installations and is replaced with a gas-fired or multi-fuel reciprocating engine. In small CHP systems the output is more likely to be geared to satisfy electricity

Figure 14.10 **Average Carbon Dioxide Emissions from Thermal Power Plants**

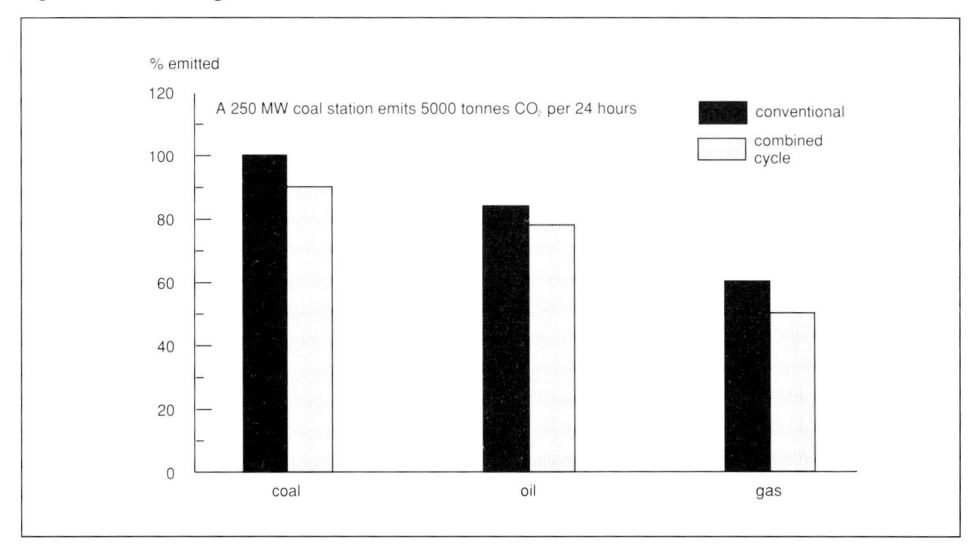

Page 182

requirements so the need to dispose of any surplus electricity should not arise; electricity can be bought in, if necessary, to satisfy peak demand periods.

Environmental Considerations

As far as power generation is concerned, environmental considerations are focused mainly on:

- *acid rain, i.e. emissions of SO_x and NO_x;*
- *the so-called greenhouse effect, i.e. emissions of CO_2, NO_x and water vapour;*
- *air quality, i.e. emissions of SO_x, NO_x and particulates;*
- *disposal of solid and liquid wastes, including radioactive material in the case of nuclear plants;*
- *'spoiling' the natural environment, particularly in the case of hydro-electric schemes;*
- *decommissioning plants at the end of their useful life; and,*
- *safety at all stages from construction through to decommissioning.*

Some of these issues, e.g. the disposal of radioactive material, solid and liquid wastes and the emission of SO_x and particulates, do not arise with gas-fired combined cycle plants, but all fossil fuels when burnt, including natural gas, produce varying amounts of CO_2, NO_x and water vapour.

The advantage that combined cycle enjoys over conventional coal and oil-fired plants is that the amounts of CO_2 and NO_x produced per kWh of electricity generated are substantially lower. Average values are shown in Figures 14.10 and 14.11; actual values will depend on the carbon content of the fuel and the efficiency of the process used. Lower values can be achieved – for example by using selective catalytic reduction to reduce NO_x emissions from coal and gas burning plants – but this is expensive in both investment and operating costs.

Table 14.1 **Typical Average Daily Activity of a 2000 MW Power Plant**

Plant Type	Fuel Consumed (tonnes)	Waste Solids (tonnes)	Waste Heat (GWh)	SO_2 Produced (tonnes)	NOx Produced (tonnes)	CO_2 Produced (tonnes)	Net Efficiency (%)
COAL (1% sulphur)							
– conventional	16 500	2 000*	76	350	50-150	39 000	38
– gasification/combined cycle	14 800	1 800*	63	negligible	10-50	35 000	43
FUEL OIL (3.5% sulphur)							
– conventional	10 200	5*	73	750	40-70	33 000	39
NATURAL GAS							
– conventional	7 800	nil	70	negligible	15-75	21 000	40
– combined cycle	6 500	nil	52	negligible	10-50	18 000	48

*Plus sulphur

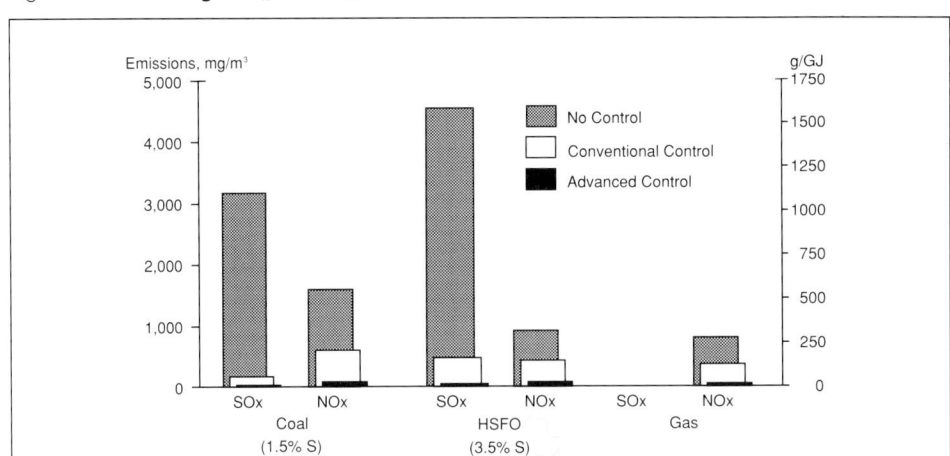

Figure 14.11 **Average NO$_x$ and SO$_x$ Emissions from Thermal Power Plants**

While up to 90 per cent of the sulphur compounds produced by conventional oil and coal-fired plants can be removed by flue gas desulphurisation equipment, it is expensive, it reduces the thermal efficiency of the power plant, and there is the added problem, and cost, of waste disposal.

Particulates from coal and oil-fired plants can now be collected and removed down to acceptable levels but again at a cost.

Decommissioning and safety issues are fairly similar for all fossil fuel plants with no system having a marked advantage over the others assuming that construction,

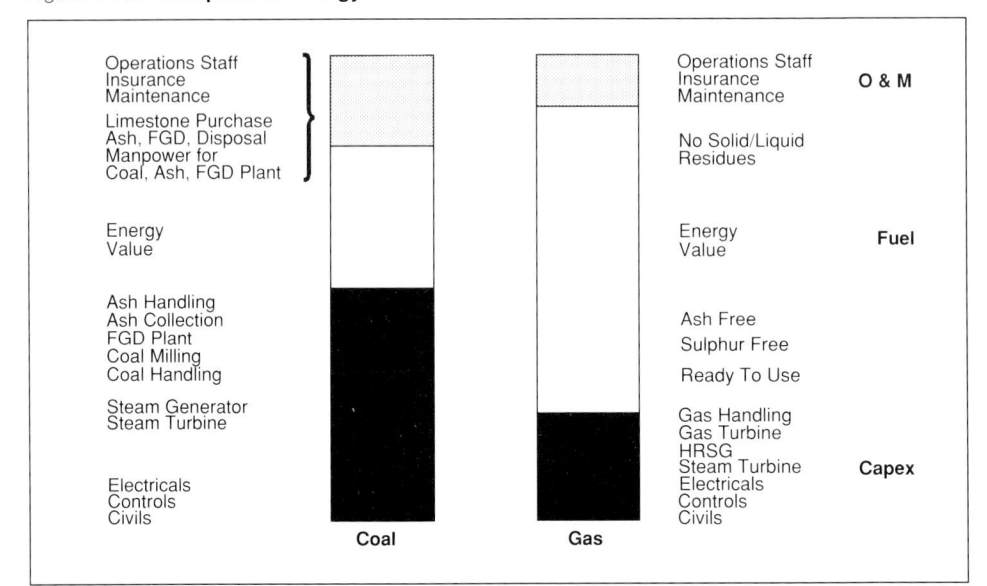

Figure 14.12 **Comparative Energy Values and Costs**

Figure 14.13 **A Combined Cycle Plant at Waidhaus Compressor Station, Germany**

Courtesy of Ruhrgas A.G.

operation and maintenance have been carried out to the necessary standards. But in virtually all other respects gas-fired combined cycle is better able to meet environmental standards, more efficiently and at a lower capital and operating cost than its coal and oil-fired competitors. However, neither the competition nor technology stand still and new systems are being developed, particularly by the coal industry, with the objective of satisfying more stringent environmental standards as they are introduced.

Relative Benefits and Gas Values

Table 14.1 summarises the main performance related advantages that a combined cycle gas turbine system offers over all other fossil fuel based technologies for generating electricity.

The ability to use natural gas in this way gives natural gas a premium or opportunity value over other fuels – see Figure 14.12 – and there are further added-value elements when its comparatively better, but more difficult to quantify, environmental performance is taken into account.

The high efficiency performance of CCGT plant is achieved at a lower specific capital cost compared to a conventional single cycle steam plant. As shown in Figure 14.12, the relative benefits of CCGT can be assessed against a conventional plant and a value for the gas obtained. The challenge then facing the gas marketer is to translate that value into a gas price that the electric utility is willing to pay.

Chapter 15

GAS CONVERSION

This chapter reviews some of the more important products that can be manufactured by synthesis, starting from natural gas, in particular from methane. It is not comprehensive and the processes outlined below are typical and not necessarily specific. This is because there are now various proven processes available to make the same products and many of these, and the catalysts they use, are proprietary. Two exceptions to this generality are Mobil's gas-to-gasoline and Shell's gas-to-middle distillates processes. These also involve proprietary catalysts but merit specific mention as they are the proven processes currently available to make these products on a commercial scale, although SASOL's coal-to-gasoline technology is planned to be used in a gas-to-gasoline project in South Africa.

Synthesis Gas

The conversion of methane, the main hydrocarbon component of natural gas, is mainly via the production of synthesis gas – otherwise known as 'syngas' – which is a mixture of carbon oxides and hydrogen. Syngas is the essential building block, the initial step in the subsequent production of a wide variety of end-products such as ammonia, urea, methanol, various liquid hydrocarbons, waxes, etc.

The two proven technologies available for the production of syngas are steam reforming and partial oxidation.

Steam Reforming

The most commonly used methane conversion process, steam reforming, could hypothetically produce, as a catalytic reaction between steam and methane, a synthesis gas with a hydrogen:carbon monoxide ratio of about 3, viz:

$$CH_4 + H_2O \rightarrow CO + 3H_2$$

However, in practice the CO shift reaction proceeds in parallel virtually to equilibrium to give:

$$CO + H_2O \rightarrow CO_2 + H_2$$

Most catalysts used for steam reforming are nickel based and an excess of steam is required to prevent carbon deposition. As these catalysts are poisoned by even small amounts of sulphur, the feed natural gas must first be processed if any sulphur compounds are present. At the outlet of the steam reformer, the raw syngas will contain CO, CO_2, H_2, unconverted methane and excess steam.

The steam reforming reaction is essentially endothermic and requires temperatures in excess of 700°C to obtain a high conversion level of the feed gas. Depending on the end-product to be made, pressures can be in excess of 20 bar.

The design of most steam reformers is based on tubes filled with the catalyst suspended in a large furnace heated mainly by radiation. The recovery of heat from the syngas and products of combustion is an important economic consideration.

Partial Oxidation

With partial oxidation the reaction to produce syngas with a hydrogen:carbon monoxide ratio of about 2 is based on oxygen, viz:

$$CH_4 + {}^1/_2 O_2 \rightarrow CO + 2H_2$$

Non-catalytic partial oxidation processes have been developed mainly by Shell and Texaco. These can produce syngas from a wide variety of hydrocarbon feedstocks ranging from natural gas to heavy oil residues. Reactors are essentially combustion chambers fed with a mixture of the hydrocarbon and oxygen. They are operated at temperatures generally in the range of 1300 to 1500°C and at pressures up to 60 bar.

An air separation plant is necessary to produce pure oxygen, and depending upon the subsequent use of the syngas, the hydrocarbon (natural gas) feed to the syngas plant must be free of sulphur (as for steam reforming) to avoid contamination of any catalysts used in subsequent processing of the syngas.

Again depending on the end-product to be made from the syngas, further processing, apart from cooling, may be necessary to produce the required syngas quality, for example to change the ratio of carbon monoxide and hydrogen, to remove unwanted components, etc.

Comparison of the relative economics of steam reforming versus partial oxidation is not helpful in general terms as this depends on the nature of the hydrocarbon feed, the quality required and subsequent use to be made of the syngas, and other considerations.

Methanol

This is a growing application for natural gas as a consequence of the increase in world demand for methanol for the chemical industry, e.g. for the production of formaldehyde, acetic acid, etc, and for blending components for automotive fuels.

The preferred syngas quality for methanol production has a hydrogen:carbon monoxide ratio of about 2 parts to one. The reactions are:

$$CO + 2H_2 \rightarrow CH_3OH$$

$$CO_2 + 3H_2 \rightarrow CH_3OH + H_2O$$

In methanol synthesis CO and CO_2 are reasonably interchangeable; in some processes carbon dioxide may be added from an outside source.

The reaction between hydrogen and carbon monoxide to produce methanol is highly exothermic, while that between carbon dioxide and hydrogen is less so producing about one-half the heat. Control of reactor temperatures at 250 to 270°C and pressures at about 100 bar, plus efficient heat recovery, are key features of plant design. Most processes use a copper-based, fixed-bed catalyst system with either tube cooling or intermediate quenching between the catalyst beds.

For most applications high quality product is required; water and other impurities will be removed by distillation.

Hydrogen

The production of hydrogen from natural gas dates back to the early 1930s. The main uses of hydrogen are for hydrogenation in the oil industry and the production of ammonia. For such purposes all CO is converted into CO_2 by shift conversion and the CO_2 is removed from the syngas by alkaline scrubbing with an amine or regenerative caustic solution. The resulting hydrogen-rich gas is then purified.

Ammonia and Urea

Natural gases containing relatively high percentages of nitrogen are advantageous for the synthesis of ammonia from nitrogen and hydrogen. The reaction is basically:

$$3H_2 + N_2 \rightarrow 2NH_3$$

The syngas feed for ammonia synthesis needs to have the correct stoichiometric ratio of hydrogen and nitrogen, also to be free of oxygen and carbon oxides to avoid deactivation of the iron-based catalyst. The process is more complex than that for

Figure 15.1 **General View of an Ammonia Plant**

Kemira Fertiliser's plant, Ince, Chester, England. Plant input 40 million ft^3/day of natural gas. Plant output 950 tonnes/day. Courtesy of Kemira Ince Ltd.

Figure 15.2 **ICI's Ammonia Plant, Severnside, Bristol, England**

In the centre are two 450 t/d process units one behind the other. In the left foreground is a gas heated reformer, next to it is an isothermal CO shift reactor. The rows of vessels to the right are part of the pressure-swing adsorption system. Courtesy of ICI Chemicals & Polymers Ltd.

methanol production requiring the addition of a secondary reformer using air, CO_2 removal and methanation units. The latter is required to react any residual carbon oxides with hydrogen to produce methane which can be tolerated within the ammonia loop synthesis process. With some processes more air has to be added than needed for the reaction in order to generate more heat in the secondary reformer.

Typically, synthesis occurs at temperatures in the range of 380 to 450°C and at pressures with current technology down to around 80 bar. Research efforts are aimed at finding catalysts which will perform at lower temperatures as these should offer higher conversion rates and better economics.

A key feature of the ammonia synthesis loop design process is gas circulation as conversion per pass over the catalyst is limited by equilibrium to about 15 per cent.

Ammonia, which is a gas at normal temperature and pressure, is used in combination with other feedstocks mainly for the production of solid fertilizers, i.e. ammonium sulphate, ammonium nitrate, ammonium phosphates, etc. However, urea, another important fertilizer, can be manufactured by combining ammonia and carbon dioxide. The reaction is:

$$2NH_3 + CO_2 \rightarrow (NH_2)_2CO + H_2O$$

In the conversion of natural gas to syngas as the feed for ammonia production, sufficient carbon dioxide is produced as a by-product which can then be used to provide the feed for the urea reaction. Thus urea can be produced from ammonia without having to purchase or manufacture another reagent to convert the ammonia into solid urea. About one tonne of ammonia produces about 1.6 tonnes of urea.

Gasoline

As mentioned earlier, unlike the other products described above for which there are many plants with different processes and catalysts in service today, there is only one commercial-scale plant in the world at present which produces gasoline from natural gas. This is located in New Zealand and is based on a fixed-bed MTG (methanol-to-gasoline) process devised by Mobil and operated in combination with ICI's low pressure methanol synthesis process. From an input of some 130 million ft^3/day it can produce nearly 15000 barrels/day of gasoline. The plant is reported to have cost, including start-up costs and interest during construction, about US$1.5 billion.

The key to this MTG process is Mobil's proprietary synthetic zeolite, ZSM-5, catalyst which limits the resulting products to the gasoline range. It can be operated at high temperatures without losing its activity and physical integrity. The total process involves the production of syngas from the natural gas feed which is then converted to methanol using an ICI process. The resultant methanol is then dehydrated to dimethylether (DME) over activated alumina catalyst. The DME then undergoes several reactions over the ZSM-5 catalyst to form light olefins (C_2 to C_5), viz:

$$2CH_3OH \rightarrow CH_3OCH_3 + H_2O \rightarrow C_2H_4 + H_2O$$

Figure 15.3 **SMDS Plant Construction Activities at Bintulu**

Courtesy of Shell MDS (Malaysia) Sdn. Bhd.

These light olefins are reacted further to form heavier olefins (C_5+) and are then rearranged into paraffins, cycloparaffins and aromatics via hydrogen transfer.

With sufficiently long residence time in the reactor, the conversion of methanol is virtually complete. The gasoline yield exceeds 80 per cent by weight, the balance of hydrocarbon products being small quantities of methane, ethane, ethylene, propanes and butanes. Gasoline quality is virtually equivalent to regular unleaded.

Other processes to convert natural gas to gasoline have been developed but none as yet have reached the commercial stage. However, as already mentioned, there are several plants in operation in South Africa which produce gasoline and other liquid products from coal.

The likelihood of more commercial scale natural gas-to-gasoline plants being built will depend to a large extent on such factors as future prices and perceived shortages of gasoline from conventional sources. This is because gas-to-gasoline processes are capital intensive and require relatively low gas prices and high gasoline prices to be economically viable.

Middle Distillates

The world's first plant to convert natural gas-to-middle distillates is now being built by Shell in partnership with Petronas, Mitsubishi and the Sarawak Government at Bintulu, Sarawak, Malaysia, and is due to come on stream in 1993. From a feed of 100 million ft^3/day, this plant will produce gasoil, kerosene, naphtha, normal paraffins and several grades of wax, totalling some 470000 tonnes per year. The acronym for the total process is SMDS (Shell Middle Distillate Synthesis).

The process involves the removal of all traces of sulphur from the feed gas with zinc oxide beds before syngas is produced by partial oxidation using the Shell Gasification Process. This gives a synthesis gas with a hydrogen:carbon monoxide ratio of about 2 which without much correction is a suitable feed for the production of middle distillates.

The next step, Heavy Paraffin Synthesis, is the heart of the process and is a modernised version of the classical Fischer-Tropsch technique. It makes use of a proprietary Shell catalyst that combines high activity with high selectivity for the desired products. In this step the synthesis gas components, hydrogen and carbon monoxide, react to form predominantly long chain paraffins which extend well into the wax range.

The next steps comprise Heavy Paraffin Conversion and various distillation units. Here naphtha, kerosene, gasoil, paraffins and various wax grades are produced. According to market demand, the long chain waxy paraffins are either left intact or, using a special Shell catalyst, broken up into molecules of desired sizes, mainly in the kerosene and gasoil range.

Butanes and lighter hydrocarbons are sent back to the process to be converted into hydrogen in the Hydrogen Manufacturing Unit. This hydrogen is predominantly fed to the Heavy Paraffin Synthesis unit in a mix with the synthesis gas and to the Heavy Paraffin Conversion unit.

In essence, the process is simple, combining conventional and well proven technologies in all but the Heavy Paraffin Synthesis step. The novelty of this step, which is also the key to the whole process, is in the use of the Shell developed and improved Fischer-Tropsch catalyst. The reactions are:

$$CH_4 + {}^1/_2 O_2 \rightarrow CO + 2H_2 \rightarrow (-CH_2-) + H_2O$$
$$\text{Heavy paraffins}$$

By adjusting recycle rates and cut-points in the distillation stage, the product slate can, within certain limits, be adjusted to market requirements – see Table 15.1. The products themselves are completely paraffinic, free from nitrogen and sulphur, and are of the highest possible quality.

Table 15.1 **SMDS Product Variability**

Product	Gasoil Mode (% mass)	Kerosene Mode (% mass)
Tops/naphtha	15	25
Kerosene	25	50
Gasoil	60	25

Figure 15.4 **Ethane Cracker: Mossmorran, Scotland**

This cracker produces 500000 tonnes/year of ethylene from a feed of about 2300 tonnes/day of ethane from the adjacent NGL fractionation plant. The former is operated by Esso Chemicals and the latter by Shell UK Exploration & Production.

The selectivity of the process in respect of the conversion into liquid hydrocarbons (C_5 + material) is 78 to 80 per cent. The overall thermal efficiency based on lower heating values of both the gas and the middle distillate products is 63 per cent.

Given the rapidly rising demand for high quality middle distillates, in particular for automotive fuels, the potential scope for efficient, cost-effective gas-to-middle distillates processes appears to be substantial. But as with virtually all gas conversion projects, economic viability to a lesser or greater extent will be conditioned by world prices for the end-products in question.

Methyl Tertiary Butyl Ether (MTBE)

Demand for MTBE is expected to grow considerably as it is used increasingly as an octane booster, as leaded motor gasoline is progressively phased out, and as a component of 'reformulated gasoline', a generic expression which applies to any gasoline whose composition has been changed to reduce exhaust emissions.

There are several processes which produce MTBE in all of which isobutylene is reacted selectively with methanol to produce MTBE. It is a relatively simple combination reaction at moderate temperatures of 60 to 100°C and gives a high yield. Strongly acidic ion-exchange resins are generally used as catalysts. The reactions are:

$$CH_2 = \underset{\underset{CH_3}{|}}{\overset{\overset{CH_3}{|}}{C}} + CH_3OH \rightarrow CH_3 - \underset{\underset{CH_3}{|}}{\overset{\overset{CH_3}{|}}{C}} - O - CH_3$$

$$\text{isobutylene} \quad \text{methanol} \qquad\qquad \text{MTBE}$$

In the past, most isobutylene originated as a by-product from petrochemical plants (steam cracking) or refineries (catalytic cracking), or manufactured by dehydrating isobutanol. However, these sources are now becoming fully utilised, and further increases in demand will have to be supplied, at least in part, from field butanes. The latter are first isomerised and then dehydrogenated, thus forming a relatively expensive source of isobutylene, but the methanol to MTBE stage is relatively low cost. With natural gas already being the principal feedstock for the production of methanol, MTBE is increasingly becoming a gas-based product.

Other Products

Other products which can be manufactured either directly from the synthesis of methane, or in combination with or as derivatives of other products, include acetylene, acetic acid, dimethyl ether, chloromethanes, hydrogen cyanide, formaldehyde, olefins – there are others. While the processes to make these products are proven, some of them offer relatively low yields and may not be economic on a commercial scale.

Another related field of natural gas conversion not discussed above but which deserves mention for completeness sake, is the conversion of some other components of natural gas, notably ethane, propane and butane. Particularly important in this respect is the conversion of ethane, propane and butane to ethylene by non-catalytic steam cracking (pyrolysis), also the catalytic dehydrogenation of propane and butane to produce propylene and, as discussed above, isobutylene. These products are essential building blocks for the modern-day petrochemicals industry.

Research and Development

Despite considerable research efforts to convert methane directly to other products by, for example, oxidative coupling or direct oxidation, conversion first to synthesis gas continues to be the only feasible commercial option. The basic problem is that because the methane molecule is very stable, most reaction products are more reactive than methane itself. In consequence, it is difficult to stop reactions before they result in complete oxidation and/or the production of elemental carbon, and control of the product slate is not easy.

Producing syngas as a first step is expensive and consumes a lot of energy, although it does yield a high quality material for further synthesis. While research continues to find means of converting methane directly, effort is also being devoted to devising more efficient catalysts and improved process designs so as to obtain higher yields at lower cost with proven technology.

Readers who require more information than has been given in the foregoing simplified account of gas conversion are advised to research the wealth of literature available on this subject.

Chapter 16

MARKET ASSESSMENT AND UTILISATION PRIORITIES

There is no definitive list of natural gas utilisation priorities that can be applied universally, nor any one methodology for assessing market prospects. In practice, account has to be taken of governmental energy and especially fuel taxation policies, the availability, cost and use of alternative energies, population densities, living habits and standards, industrial activity, balance of payments issues, ambient temperature ranges, geographic and logistical factors, environmental concerns and so on.

All that can be offered in abstract is to indicate utilisation priorities that have been followed in a number of countries where there has been a large measure of freedom of choice in developing a new market. Similarly, to suggest how a market can be surveyed and assessed to ascertain if the marketing of gas is likely to be viable. Although they are discussed separately, in reality utilisation priorities and market assessment are inter-related and react on each other.

For the purposes of this discussion, it is assumed that a substantial reserve of good quality gas has been discovered and that all indications point to it being technically exploitable.

Before listing possible utilisation priorities the following generalisations should be borne in mind. First, is the need for a rapid build up to achieve economies of scale at an early stage because of the high incidence of fixed costs of gas infrastructures. Second, but no less important, is the need to operate the infrastructure at the highest possible load factor to minimise unit costs – see also Chapter 10.

Other considerations which may apply are, for example, taking advantage of any available infrastructure such as converting an existing low calorific value manufactured gas distribution system to high calorific value natural gas, or giving preference to the development of large base-load outlets using conventional technology rather than to applications which require more complex technology and specialist expertise. The ability to finance new gas projects is a further consideration, as will be the need to attract and involve experienced operators in one, several or all phases of a gas venture. Fiscal terms and the ability for foreign operators/partners to remit earnings will undoubtedly condition the degree to which they may be willing to become involved in helping to get a new gas venture off the ground quickly and successfully.

These and many other factors need to be taken into account when deciding which priorities for gas utilisation are appropriate for any particular country.

First Priority: Maximise Local Uses

The objectives here are to reduce the need for oil (and/or coal) imports with consequential foreign exchange savings, or to release more oil for export as this usually has a greater revenue earning potential than gas on an equivalent energy basis. Gas also offers environmental benefits over most other energies. The suggested order of development for local uses is:

1. LARGE INDUSTRIES and POWER PLANTS, essentially steam-raising and other bulk heating uses.

Main advantages:

- *gives economies of scale (large individual offtakes)*
- *offers good load factor*
- *can avoid costly treatment facilities as gas quality is not usually critical*
- *provides basic gas infrastructure for subsequent expansion*
- *no storage on users' premises or waste disposal problems*
- *uses conventional technology*
- *improves the environment; i.e. no SO_x and less NO_x and CO_2 emissions, no particulates, etc., compared with coal and most oil products.*

Main disadvantages:

- *possible low value in competition with coal, fuel oil, hydro, etc.*
- *load factor may be low if the end-consumer installs dual-fired capability or buys gas on an interruptible basis.*

2. SMALL/MEDIUM INDUSTRIES and COMMERCIAL users.

Main advantages:

- *higher values than steam-raising and similar basic heating uses*
- *high load factor offtakes, unless space heating is involved*
- *no storage on users' premises or waste disposal problems*
- *good volume potential, but slower build-up than for steam raising and like uses*
- *ease of control/flexibility at point of use*
- *improves the environment*

Main disadvantages:

- *need for treatment facilities as gas quality is usually important*
- *cost and operation of a more complex distribution system*
- *more sophisticated technology at point of use than steam raising*
- *need for service/maintenance back-up support*

3. RESIDENTIAL

Main advantages:

- *high gas values in 'free market' economies*
- *high load factor, unless space heating is involved*
- *no storage on users' premises*
- *good volume potential if space heating is involved, but usually a slow build-up unless there is an existing town gas grid*
- *ease of control/flexibility at point of use*
- *improves the environment*

Main disadvantages:

- *need for treatment facilities as satisfactory/constant gas quality is essential*
- *high cost of gas distribution system unless there is an existing town gas grid*
- *substantial service/maintenance back-up support essential*
- *relatively low average offtakes per connection even if space heating is involved in which case the load factor will be low.*

Second Priority: Exports as Gas

Exports generate foreign exchange earnings and utilise gas which is either surplus to long term local needs or for which no local market can be developed. However, exports usually require relatively large reserves after allowing for any local uses.

By PIPELINE overland or subsea to neighbouring countries.

Main advantages:

- *where feasible/practicable, pipelines are usually quicker and easier to construct and operate than LNG projects*
- *conventional technology unless deep-water crossings are involved*
- *offtake prospects in other countries that may have to be crossed en-route to the ultimate market*
- *substantial revenue earning potential if volumes are large*

Some constraints:

- *large offtake volumes are usually necessary for exports to be economic*
- *potential supply security problems if several countries are involved*
- *maintenance/operation of the system in foreign countries may be difficult*
- *net-backs diminish with reducing volumes and/or increasing distances*
- *contractual and corporate structure complexities if several countries are involved*

As LNG across oceans where pipelines are impractical.

Main advantages:

- *direct supply/contractual relationship between seller and buyer(s)*
- *greater local employment prospects than for pipeline gas exports*
- *technology transfer potential which can be particularly important for developing countries*
- *expansion is physically easier than pipeline supplies by adding more liquefaction capacity/ships*
- *scope for producing countries (as commercial partners) to participate outside national boundaries, i.e. in shipping and/or trading*

Some constraints:

- *sophisticated technology*
- *capital intensive with long lead times*
- *need for regular, high offtakes to be economic*
- *some energy is 'lost' in the liquefaction process*
- *net-backs diminish with reducing volumes and/or increasing distances*

Third Priority: Conversion to other Products

The principal objective of gas conversion projects is to produce products such as ammonia, urea, MTBE, methanol, middle distillates, etc., for local consumption, or for export, or for both. In some instances where gas reserves are too small to support a pipeline gas or LNG project, and where there is no local gas market of consequence, conversion projects may be the only viable option.

Some considerations:

- *volumes of gas utilised in conversion are modest compared with most pipeline gas and LNG projects*
- *low into-plant gas prices are usually needed to make conversion projects viable if the prices of alternative feedstocks, notably oil products, are also relatively low*
- *capital intensive with long lead times*

- *sophisticated technology is involved in most instances*
- *energy is 'lost' in the conversion process*
- *some products may become over-supplied and therefore difficult to market*
- *export prices/net-backs will be influenced by competitors' activities and world prices*
- *projects may need 'under-writing' by governments to succeed*
- *lower revenue earning potential in absolute terms than pipeline gas or LNG*

If reserves are far greater than the local market can ever expect to utilise itself, scope may exist to develop several or all of the above possibilities.

The foregoing priorities will also be generally applicable where imports of pipeline gas or LNG are being considered.

A Methodology for Market Assessment

Just as there is no definitive list of utilisation priorities, equally there is no definitive methodology for market assessment. All that can be said is that the methodology described below has been used successfully on a number of occasions. If it appears to be over-elaborate and cumbersome, then it should be remembered that even relatively simple marketing structures can cost several hundred million dollars, while a large and complex structure, together with the necessary upstream production facilities, can amount to several billion dollars.

A market assessment may be undertaken for a fee by an experienced gas entity at the request of government, by a producer who needs to convince himself that a worthwhile market exists before investing in expensive production facilities and/or to evaluate for himself the likely level of ex-field prices he can expect to achieve, or by the prospective investing entity which may be the producer, a third party, or some combination thereof. A prospective exporter of pipeline gas or LNG may also wish to assess the market in the target importing country, particularly if it is a greenfield market, for much the same reasons.

Assessing whether a particular potential customer will actually convert to natural gas involves a variety of factors. Basically, one can be fairly certain that he will not convert unless there is a positive attraction to do so. Frequently, the attraction will be of a short and/or long term economic nature. However, it can also be because conversion to gas improves the quality and therefore the value of his product. But if the economic incentive is small it may be insufficient to persuade the more conservative potential gas user to take the risk of converting to what to him is a new fuel of untried reliability. In making a decision he is also likely to be influenced strongly by the proportion of his total annual costs which are energy costs. Generally, the higher the proportion his energy bill is of his total costs the smaller the saving which needs to be offered before he will convert. Although in the final analysis the size of the gas market will be determined largely by the price competitiveness of gas relative to alternate fuels already being used, or that are planned to be introduced, in practice there are other factors to be considered.

Assessing the Large Users' Market

Starting with the industrial and power generation markets, as is usually the case for the reasons given earlier, the market analyst will first need to gather as much information as he can on the location, the nature and size of the business, and the type(s) of fuel(s) used by the larger industrial enterprises and power plants in the target market area or areas. Part of this information is likely to be available from published statistics, but these are seldom up-to-date, frequently incomplete, not necessarily precise, and are unlikely to give much information, if any, on fuel consumption patterns and the fuel prices paid by individual consumers. However, having gathered such information as is available, the analyst is then in a better position to start the next phase of his study.

This is to embark on a factory-by-factory survey, or at least a good representative sample of each type of industrial activity, using a combination of written questionnaires and personal interviews to obtain details of fuel usage, prices, etc., and, if possible, indications of expected future energy consumption levels. All major fuel users, especially power plants, should be visited. This is time-consuming work but necessary in order to estimate the total industrial fuel demand which is potentially vulnerable to penetration by gas. Note will also be taken of firm plans to build new industrial complexes and power plants and the fuels they intend to use. In this regard, it may be appropriate to persuade electric utilities to adopt combined cycle gas turbine technology for new power plants thus creating a large ready-made outlet for gas. The assistance of government departments and local planning authorities can be vital in these matters.

This total potential is not, of course, that which can be achieved by natural gas. In reality the scope for gas will depend on such factors as price, conversion costs, existing contractual commitments, etc. Accordingly, an assessment of the 'achievable volume' has to be made using a vulnerability calculation which compares the fuel bill each consumer would incur if he stayed with his existing fuel over some chosen comparison period, say three years, with that which he would incur if he switched to natural gas. This comparison should take into account such factors as differences in thermal efficiencies and operating costs, conversion costs and, if necessary, some form of 'incentive allowance', i.e. some saving of the customer's existing fuel bill which makes the switching to gas attractive to him. Where an incentive allowance is considered necessary, it would only apply until conversion costs are recovered, thereafter the customer should expect to pay close to the full market value price as set by the going price(s) of alternative fuel(s).

By changing the gas price up and down the number of customers who are potential candidates for conversion to natural gas will vary. And by reiteration using a suitably designed computer program, a gas price can be calculated which maximises revenues while at the same time offering that group of customers sufficient incentive to convert to gas. This base case assessment of the maximum revenue generating price assumes that the same price would apply to all customers irrespective of their offtake and geographic location, whereas in practice it may be possible to fine-tune prices to take account of these factors so as to maximise revenues still further. Further complications can include individual customers who use different fuels at different prices for different applications, load factor variations, quantity discounts, different tax regimes and/or government subsidies, etc. But there is a limit to the degree of sophistication and detail that can be considered and built-in by the market analyst and this will vary from one situation to another.

Once the potential large users' gas market has been evaluated in volumetric, geographic and price terms, the next main step is to assess the type of pipeline system and its cost that would be required to supply gas to the identified prospective customers. Here again a computer program is required to examine on an iterative basis different routes and configurations, pipe diameters, operating pressures, the possible need for compressor stations if long distances are involved, the nature of the terrain to be crossed, the number and location of branch lines required, etc.

The usual procedure would be to evaluate the optimum system configuration required for a selected horizon year. When that has been done the analyst then returns to the first year and selects from the horizon year configuration that part of the system which would be required to meet initial demand. This process can then be repeated for subsequent years so as to gain a picture of how the system would be developed over time. Calculations are made of capital investment and operating costs for each main section of the system as well as for the system as a whole.

The third step, using the data gleaned from the first two, is to analyse the economic viability of the proposed venture and, if necessary/appropriate, for parts of it, e.g. for those customer(s) who may be remotely located from the expected routing of the main system, or those whose business may be crucial in order to realise the necessary economies of scale, etc. A vital consideration here is, of course, the cost of gas, particularly if the gas supplier is a separate entity who will naturally wish to obtain the maximum value he can for the gas he supplies. But irrespective of who the gas supplier may be, obviously a firm indication, if not a definitive price, is necessary. The alternative approach is that having assessed the cost of transporting and distributing the gas to the target market, the proposed marketing entity is then in a position to tell the gas supplier what he can afford to pay for the gas.

The economic assessment will not only include the capital and operating costs involved, but the desired rate of return, present value cash surplus, pay back time and other financial and profitability criteria as discussed in Chapter 17. Political risks, operatorship, technical and other back-up support services required, etc., also need to be taken into account.

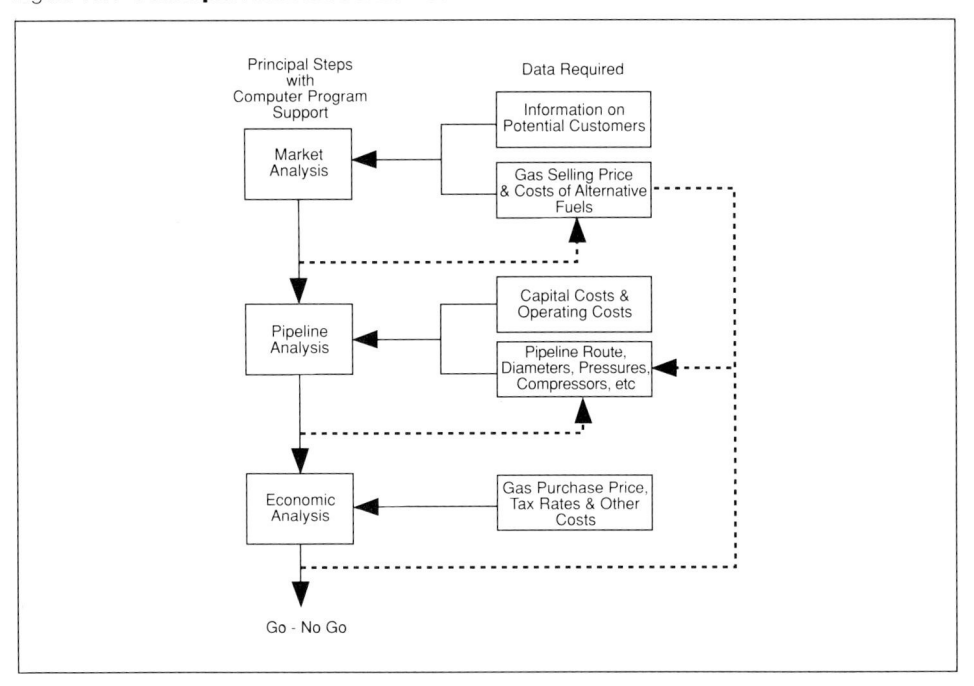

Figure 16.1 **Principal Activities involved in Market Assessment**

```
Principal Steps                Data Required
    with
Computer Program
   Support              ┌──────────────────────┐
                        │ Information on       │
                        │ Potential Customers  │
   ┌──────────┐         └──────────────────────┘
   │ Market   │◄────────┤
   │ Analysis │         ┌──────────────────────┐
   └────┬─────┘         │ Gas Selling Price    │
        │               │ & Costs of Alternative│── ─ ┐
        │               │ Fuels                │     ┊
        │               └──────────────────────┘     ┊
        ▼                                            ┊
                        ┌──────────────────────┐     ┊
                        │ Capital Costs &      │     ┊
                        │ Operating Costs      │     ┊
   ┌──────────┐         └──────────────────────┘     ┊
   │ Pipeline │◄────────┤                            ┊
   │ Analysis │         ┌──────────────────────┐     ┊
   └────┬─────┘         │ Pipeline Route,      │◄─ ─ ┤
        │               │ Diameters, Pressures,│     ┊
        │               │ Compressors, etc     │     ┊
        │               └──────────────────────┘     ┊
        ▼                           ▲                ┊
                                                     ┊
   ┌──────────┐         ┌──────────────────────┐     ┊
   │ Economic │◄────────┤ Gas Purchase Price,  │     ┊
   │ Analysis │         │ Tax Rates & Other    │     ┊
   └────┬─────┘         │ Costs                │     ┊
        │               └──────────────────────┘     ┊
        ▼                                            ┊
                                                     
   Go - No Go
```

In the worst situation it may be that the outcome of all this work is that the returns are too low, or even negative, to justify developing the market. And in marginal cases it may be necessary to re-examine the whole project with the aim of trying to establish a different, more robust marketing plan.

But whatever the outcome an analysis along the above lines cannot be definitive, and unless the result is sufficiently robust to cover most eventualities the decision to go ahead may depend on the perceived balance of risks and rewards.

The Small Industrial and Commercial Markets

These markets can be important outlets for gas, but to avoid unnecessary repetition they can be discussed briefly.

Assuming that the conclusion from the foregoing analysis is to proceed with development of the large users' market, the natural consequence would be to examine the additional scope for selling gas to smaller industrial and commercial users. If a preliminary scouting study indicates that this could be of a relatively substantial nature, then the analyst may need to repeat the process described above.

The outcome of this assessment may necessitate some modifications to the size, configuration and routing of the proposed pipeline system. Other factors such as the achievable price could also be different. However, in most situations it is unlikely that small industrial and commercial users could justify a market development venture on their own, as large user offtakes to achieve economies of scale at an early stage will most probably remain an essential prerequisite for virtually any greenfield gas marketing venture.

The Residential Market: General Aspects

The residential market poses its own special problems. If there already exists a manufactured gas distribution grid system in one or several big towns and cities as was the case, for example, in virtually all European countries before the advent of natural gas, then this may be the first market sector to be developed. In such circumstances, the analyst's job is much easier as there will already exist a wealth of information on gas usage, prices, numbers of customers and so on. Moreover, the prospective natural gas marketer will most probably be welcomed with open arms by the gas utility concerned anxious to convert his system to natural gas with all the advantages that it offers. The analyst's prime task here will be to establish in conjunction with the utility concerned how quickly and by how much the existing system can be expanded, the price levels that need to be set to cover the cost of conversion and to promote increased consumption. The scope for space heating will be an important consideration in this regard.

Where no manufactured gas system exists and where ambient temperatures are too high for space heating, then development of the residential market just for cooking and water heating is unlikely to be economically viable in isolation and may be postponed until such time as it can be developed on the back of a well established industrial business. There may be, of course, some exceptions to this generality where, for example, government is anxious for social/political reasons to develop the residential gas market and to subsidise in some way the substantial infrastructure costs involved. Building gas distribution grids in new townships and suburbs at the same time as other services are being installed can facilitate penetration of the residential market at a lower cost and with no inconvenience to the public which would be unavoidable when constructing a new grid in a densely populated area.

Assessing a New Residential Market

Assuming that the residential market merits assessment, then much the same basic steps as discussed earlier for the larger users' market would be undertaken, but with some additions and modifications.

In cold or temperate climates a key consideration will be the scope for space heating. A threshold temperature, i.e. the ambient temperature at which space heating is not required, will need to be established. This may be somewhere between say 15 and 18°C depending upon living habits and customs, the level of comfort generally desired, the nature of the dwellings and types of construction materials commonly used, standards of insulation applied, etc. The threshold temperature can then be compared with historical records of the yearly ranges and durations of ambient temperatures to arrive at the number of days in a typical average year, also for the coldest year that is likely to occur, when space heating is likely to be needed. From this it is then possible to calculate the load factor and the capacity of the distribution system that would be needed to supply a given number of consumers once the latter has been established.

If there are likely to be a few very cold days in a typical year then provision of peak shaving facilities, e.g. propane-air plants, may need to be considered to avoid unnecessary over-sizing of the grid.

Meanwhile, data will need to be collected by questionnaire and personal interviews from a representative range of householders and landlords on the existing fuels they use for cooking, water heating and space heating and the prices they pay for these fuels in the target market area. By applying the same methodology as described for the industrial market, a maximum achievable gas price can be calculated and hence the number of customers that may be expected to switch to gas at that price. Analyses of the distribution system needed and resultant economics would follow.

Other factors that need to be taken into account are that whereas prospective industrial customers may be prepared to switch to gas relatively quickly, development of the residential market may take much longer if prospective consumers are not accustomed to using a gaseous fuel and/or are reluctant to replace their existing appliances unless some incentive is offered. Moreover, building a new grid in congested built-up areas can be a slow process, while the creation and training of the necessary support services and personnel takes time.

Chapter 17

FINANCIAL AND ECONOMIC ASPECTS OF PROJECT EVALUATION

Every business enterprise will have its own financial and economic methods and criteria to evaluate the expected profitability of a prospective capital project. This chapter discusses certain evaluation methods for assessing profitability used by many business enterprises. There are other evaluation methods and for some entities profitability may not necessarily be the prime objective and can be subordinate to other considerations of a political, national, strategic, environmental or similar nature. In some instances, e.g. for governments, such considerations may be of over-riding importance. They are not discussed any further here. The focus of this discussion is how to measure profitability in relation to the capital investment to be used and the risks involved.

Screening and Ranking

These are two principal aspects in evaluating the expected profitability of a capital project. SCREENING is the process of determining whether a particular project is likely to produce an acceptable return or not. This is usually done by setting a minimum return and accepting any project that is expected to achieve this return.

Screening assumes that there are no constraints on the financial resources available to the investor and that all projects being considered are not mutually exclusive. This does not imply that the investor has unlimited funds, but that the combined value of the prospective projects that meet the chosen screening criteria does not exceed the total funds available for investment. If it does, then obviously a higher screening rate will need to be applied.

In many multi-activity enterprises the minimum rate of return for project screening purposes will often be set at a higher level than the historic rate of return actually achieved on the enterprise's total assets. This is partly because these total assets will probably include past investments which did not achieve the screening criteria but were justified on grounds of safety issues, the need to meet mandatory legal requirements or environmental standards, unquantifiable exploration in new areas, etc., and partly because of possible 'appraisal optimism' in evaluating projects. Nevertheless, the past performance of similar types of projects to those now being screened provides a useful indicator in helping to set the chosen screening rate.

RANKING is usually used when alternative investment opportunities are mutually exclusive. For example, it may not be possible for any one enterprise to undertake a pipeline gas project, an LNG project and a gas conversion project at more or less the same time because of limitations in the resources – capital, human, etc. – available to it. Where a set of projects is mutually exclusive, the set needs to be ranked separately and only the preferred project included in the overall ranking.

Measurements of Profitability

There are three main methods of measuring or assessing the estimated future profitability of a project. They are:

- *payback or payout*
- *accounting rate of return*
- *discounted cash flow (internal rate of return)*

and there are complementary subsets to some of these. Each of these main methods has its advantages and shortcomings as discussed in subsequent sections. As they all concern cash flows in one way or another, the main components of a cash flow are listed below not all of which need necessarily apply in every case.

Cash Flow Components

The main components of a CASH OUTFLOW are:

- *CAPITAL EXPENDITURE. The cost of fixed assets with a working life exceeding one year which are capitalised and depreciated in the books of accounts, e.g. pipelines, plants, ships, buildings, etc.*
- *INITIAL REVENUE EXPENSES. Costs which are not capitalised but are directly associated with a capital expenditure project, e.g. formation costs, the running costs of a plant prior to start-up, etc.*
- *WORKING CAPITAL. Stocks of gas, materials, etc., which must be held to keep the fixed assets in use, the credit given to customers (less the credit given by suppliers) and working cash balances.*
- *OPERATING COSTS. All running expenses including wages, rents, fuel, overheads, etc., but not depreciation.*
- *TAXATION. Tax payments on income, royalties and other fiscal charges.*
- *FINANCIAL COMMITMENTS. Repayments of capital and interest on borrowed funds, ship charters/leasing arrangements, etc.*

The main components of a CASH INFLOW are:

- *GROSS INCOME. Sales proceeds and any cost savings that may be realised.*
- *RETURN OF WORKING CAPITAL. Any reductions in working capital over the life of the project and the settlement of outstanding credit at the end of the project's life.*

- RESIDUAL VALUE. *Proceeds from the sale of land and worn out or partially worn out equipment and other assets. Residual value is not necessarily the same as book value.*

For any year of the project's life, the difference between the sum of the cash out and cash in components will determine whether there is an overall cash outflow (deficit) or cash inflow (surplus) for the year in question.

Payback (or payout)

The purpose of this measurement is to indicate how long money remains 'at risk' in a project. It is the number of years which must elapse before the initial cost of the project, including working capital, is returned in the form of cash, i.e. the time it takes to produce a cumulative cash outflow of zero.

The payout period starts to run when assets have been constructed and are ready for use, even though there may be a delay before any income is earned. It is common practice to add half construction time to allow for the cash outflow during this period.

Payout time on a present value basis can be calculated in the same way, taking the 'present value' instead of the undiscounted amounts of the cash flow.

Figure 17.1 **Illustrative Project Paybacks**

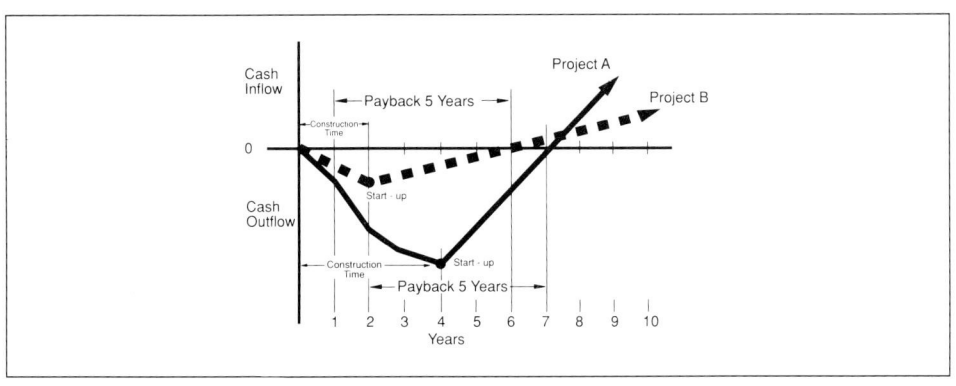

Figure 17.1 illustrates how two projects with quite different capital investment requirements and cash generating expectations can have the same payback period, in this example 5 years. Moreover, although construction started in the same year, the payback of project B is one calendar year sooner than project A and has less capital risk in absolute terms, but less cash generating potential.

Payback is simple to apply and understand and provides a measure of the risk of the project by indicating how long the initial investment remains outstanding. Its disadvantages are that it does not take into account the resources used, nor show the overall profitability of the project or the time value of money. It is a useful and meaningful evaluation tool but it should not be used in isolation.

Accounting Rate of Return

This relates the net income of the project, as determined by the normal accounting procedures of the enterprise concerned, to the accounting value of the assets employed to generate the income. Its advantages over payback are that it does recognise the resources used and it reflects the profitability of the whole project. It also shows how the project will actually appear in the financial statements of the enterprise.

However, it still retains some of the disadvantages of the payback method and introduces some shortcomings of its own. It relies on accounting conventions such as depreciation and the difference between capital and revenue which may not be well understood, may vary over time, and may be subject to manipulation. It does not take account of the time value of money. Moreover, if based on historical cost accounting, or a forecast of how the project will appear in future accounts based on it, it will tend to show an increasing return over time as the asset base is depreciated. But like payback it is meaningful and the end result is easily understood.

Discounted Cash Flow

This method overcomes most of the shortcomings of the previous two. It measures profitability by weighing each element of the cash flow according to its size and placement in time. As discussed below, the answer can be expressed as an 'internal rate of return', or a 'cumulative present value cash surplus or deficit', or both.

Clearly, discounting is the reverse of compounding. For example, if $100 invested in a risk free security at 10% per annum is worth $161 at the end of five years, the 'present value' of $161 in five year's time is $100 now. This is not the effect of inflation, since it would apply if there was no inflation and the $100 was invested to earn interest at that rate.

The reason for discounting is to enable cash inflows and outflows that occur at different times during the life of a project to be expressed in a common denominator of value, i.e. it allows for the time value of the cash flows involved. It also enables projects with different cash flow patterns to be assessed on a comparable basis.

Discounted cash flow (DCF) evaluations can be used in two ways, to show either the net present value (NPV) or the internal rate of return (IRR).

Net Present Value

Calculating the NPV of a series of cash outflows (the investment) and inflows expected over the life of the project will establish the NPV of the project as a whole – Table 17.1 illustrates the result of such a calculation. Assuming the NPV is positive, it represents the additional wealth that will be created for the investor. Conversely, if the NPV is negative at the chosen discount rate, the project will reduce the wealth of the investor. Obviously, the result will depend not only on the size and timing of the cash flow, but also on the discount rate chosen by the enterprise in question.

Table 17.1 **Illustrative Summary of Net Present Value Calculation**

Year	Annual Cash Inflow/ (Outflow) $ million		Discount Factor	Annual Present Value Surplus/ (Deficit) $ million	Notes and Assumptions
0	(100)		1.00	(100)	
1	(500)		0.87	(435)	No tax payable
2	(800)	(2700)	0.76	(608)	
3	(800)		0.66	(528)	
4	(500)		0.57	(285)	
5	400		0.50	200	Start up year
6	600		0.43	258	Assumed tax holiday years
7	800	3000	0.38	304	
8	700		0.33	231	
9	500		0.28	140	Payback year
10	500		0.25	125	
11	500		0.21	105	Other assumptions are same selling price throughout and no change in sales volumes once plateau level reached after build up period
12	500		0.18	90	
13	500		0.16	80	
14	500		0.14	70	
15	500		0.12	60	
16	500		0.11	55	
17	500		0.09	45	
18	500		0.08	40	
19	500		0.07	35	
20	500		0.06	30	This illustrative calculation would yield an earning power of about 15% in real terms
21	500		0.05	25	
22	500		0.05	25	
23	500		0.04	20	
24	500		0.03	15	
25	500		0.03	15	
Total Net Present Value Surplus				12	

Internal Rate of Return (or earning power)

This is defined as the discount rate which, when applied to a series of cash flows, produces a zero net present value. It can be calculated from any series of cash flows and represents the effective rate of return provided by those cash flows. IRR is often known as the 'earning power' of the project.

IRR, like NPV, overcomes the main shortcomings of the other evaluation methods in that it takes into account the time value of money. Being expressed as a percentage, it is easily understood and can be readily compared with the cost of capital of an enterprise. However, what IRR or earning power does not show is the monetary scale of the future cash/dividend flow or the duration of a project.

In practice, many enterprises when evaluating a major project will use several or all of the above-mentioned methods. Each has its own useful contribution to help the decision takers to decide whether or not to proceed with the project.

Inflation

It is obviously necessary to consider the effect of inflation on individual cash flow items. This can be done by starting with estimates of various cash flow items in terms of the price level applying in the base year. By applying to each of them separate assumptions about expected future price levels, a net cash flow can be calculated in 'money of the day' terms. This can then be expressed in 'real terms' by applying a general inflation index or deflator. Care needs to be exercised on how future tax paid in money of the day is treated.

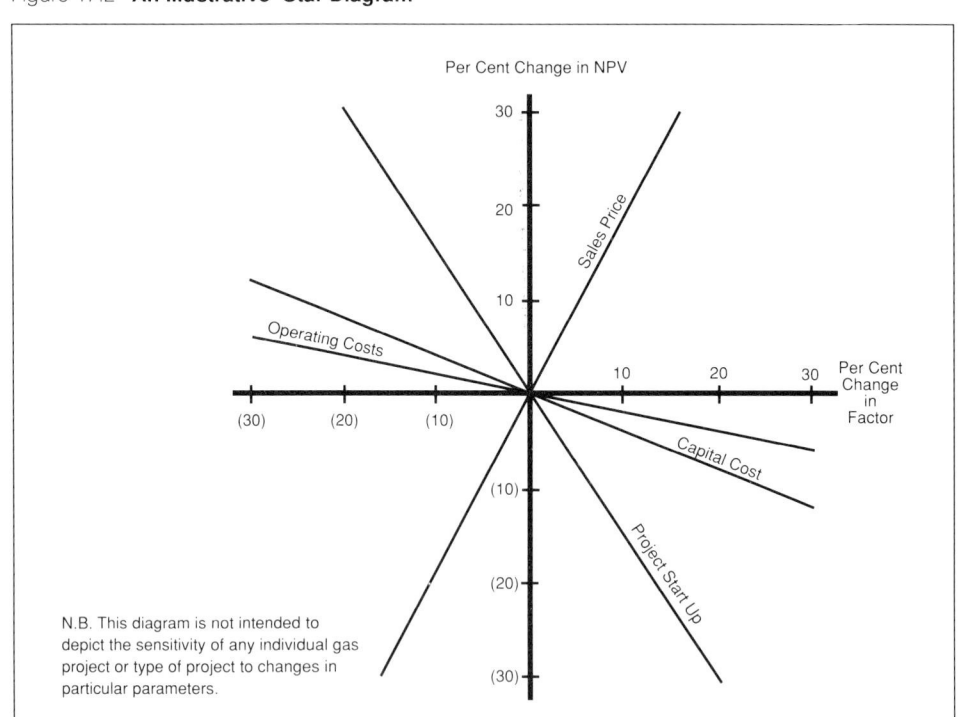

Figure 17.2 **An Illustrative 'Star Diagram'**

N.B. This diagram is not intended to depict the sensitivity of any individual gas project or type of project to changes in particular parameters.

Risk Assessment

While the project analyst will seek to obtain the best possible expert advice on all the technical, financial, fiscal, commercial and other elements he builds into his evaluations, there is still the risk that some elements may not materialise as expected. For example, completion of the project may be earlier or delayed, capital and operating costs may change, fiscal charges may alter, prices may fluctuate up or down beyond that allowed for in the basic evaluation and so on.

Any one or any combination of such changes can improve or reduce the expected profitability of the project. Inevitably, the focus will be on the possibility that the project might be less profitable than forecast, i.e. the downside risks, although the upside potential is obviously of interest, particularly if very conservative assumptions were used in the base case.

The usual way of dealing with risk assessment is to conduct a number of sensitivity tests to assess the project's robustness to cope with the consequences of various adverse changes. The approach is to change each of the key parameters in turn and to recalculate the NPV and IRR consequent upon each change. With the aid of computers this can be done with great speed and accuracy by the project analyst. If necessary, very many 'what if' versions of the base case can be computed. The results of these sensitivity tests can be depicted by using a 'star diagram' – see Figure 17.2 – which not only demonstrates the sensitivity of a project to changes in each parameter, but enables intermediate values to be read off.

While none of this work will necessarily reduce the risks inherent in a project, by being aware of those risks and their consequences it may be possible to choose a course of action which minimises the risks and to prepare contingency plans for their mitigation.

One important risk which does not lend itself readily to meaningful quantification is political risk. Where this is perceived to be very high, the project may be abandoned or postponed to await better times. In other cases where the risk is not so great, management may still decide to go ahead provided that the project is likely to be very profitable and is very robust in other respects. Alternatively, political and other risks of this nature can be taken into account by selecting a discount rate higher than would otherwise be the case.

Financing Considerations

The cost of many large gas projects is so great that even a major corporation, or a consortium of major companies, will have to raise loan finance for a significant proportion of the money to build the project and to provide working capital.

Sometimes loan finance can only be obtained if the investing company or companies, or their parent companies, are prepared to guarantee repayment of the loan should the project fail. The lender may take security over certain specific other assets belonging to the borrower – a 'fixed charge' – or take a 'floating charge' over the generality of the assets of the borrower. In other words, if the project fails to generate sufficient income to repay the loan and interest, the lender can force the sale of other assets to repay the monies owing to him. The risk to the borrower if the project should fail is that this could damage the rest of his business.

Project financing, as distinct from other means of financing a project, is where the loan is secured only against the project's assets and cash flow. Thus if the project fails, the borrower's loss is limited to his investment in the project itself and other existing business would be unaffected. However, there are also some drawbacks. For example, project specific loans tend to be more time-consuming, expensive and complex to arrange, they may require the release of sensitive information to various third parties, and may involve some loss of flexibility of a financial and/or operational nature, i.e. lenders may seek to constrain certain actions by the borrower which could reduce the robustness of the project's cash flows.

If the project is financed on a fully non-recourse basis, the lenders will pay very careful attention to the size of the equity participation and to the economic feasibility of the project to assure themselves that their risks are minimised. Important risks from their point of view, indeed for the borrower as well, are not only that the project is completed on time, but also lives up to expectations in terms of deliveries, capital and operating costs, etc.

In practice, the availability of project finance will depend on a variety of factors such as the technical, commercial and financial reputations of the project participants and the competition prevailing in financial markets. In general, and despite some of the reservations mentioned above, non-recourse project financing is usually the preferred option.

Where loans, on whatever basis they may be secured, are used to finance a project, the returns to shareholders can be different from the returns of the project on an 'ungeared' basis. This is because the interest rate charged by the lender may differ from the rate of return of the project, and because the lender will have to be paid regardless of whether the project makes the expected return.

While projects will be evaluated on both a geared and ungeared basis using the evaluation methods described above, the decision as to whether or not to proceed with a project will almost invariably be taken on the ungeared evaluation which gives a better indication of the financial strength of the project. This is because the returns to shareholders will be 'geared-up' assuming the project return exceeds the rate of interest payable on loans to finance it.

Other Considerations

The foregoing is a concise and simplified resume of what is in fact a complex and sophisticated process. Project evaluation may have to be undertaken many times as new information becomes available before the investment decision stage is reached. And assuming this is a positive decision, further evaluations will be made as construction proceeds and various negotiations with customers, banks, governments, contractors, etc., take place and perhaps change some of the earlier assumptions. While this is not an easy task, the mechanics of evaluation are greatly helped by using the proprietary software now available.

Most gas projects are of many years duration and involve major investments, in some cases amounting to several billion dollars. In addition, several millions of dollars often have to be spent on various studies to assess project viability before a decision to proceed any further can be taken. In some instances part of this expenditure may not be recoverable and may have to be absorbed in-house. Returns on gas projects usually need to be higher than many other energy projects, not only because of the commercial, technical and political risks involved, but also because almost all the capital investment required has to be sunk into the project before one dollar of sales revenue can be generated. In this regard, the present value approach is more difficult to satisfy than it is for some other energy projects where the investment is spread piecemeal over several years and sales revenue is generated much sooner. Nevertheless, that having been said, the cash generating capability of a large gas project should not be 'under-valued' given its magnitude in absolute terms, continuity and longevity.

Chapter 18

MARKET STRUCTURES AND CONTRACTS

As will become apparent, it is convenient and appropriate to address these two separate but related aspects of the business in one chapter.

Pipeline Gas Market Structures

Traditionally, the classic pipeline gas market structure has consisted of the three main segments or activities described below – see also Figure 18.1. However, as discussed later, in more recent years this classic structure has become increasingly diversified in terms of corporate activities and responsibilities.

GAS PRODUCER. The entity, or a consortium of entities, which may be privately-owned or state-owned companies, or some combination thereof, which is responsible for producing the gas and in most instances for processing the gas it produces to an acceptable marketable quality. For the purposes of this chapter, the producing entity is referred to as the 'seller'. The producer usually sells its gas ex-field or ex-processing plant, or in the case of offshore production from a reception terminal at or close to the beach landing point.

TRANSMISSION COMPANY. The entity, which may be privately-owned or state-owned, or some combination thereof, which purchases gas from the producer and transports the gas by pipeline at high pressure selling it at the 'city-gate' to gas distributors. Such entities are referred to hereafter as the 'buyer'. Frequently, transmission companies also sell gas directly to large volume industrial customers,

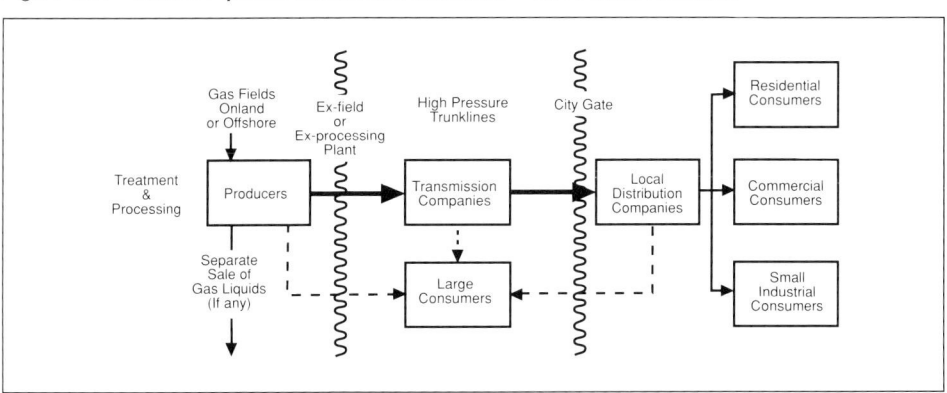

Figure 18.1 **Classic Pipeline Gas Market Structure – Three Main Activities**

LOCAL DISTRIBUTION COMPANIES (LDCs). Entities which may be municipally-owned, state-owned or privately-owned, which distribute gas at medium and low pressures selling it to end-consumers, notably, but not necessarily exclusively, residential, commercial and relatively small volume industrial customers.

As gas distribution involves supplying residential households, shops, offices, hotels, schools and the like mainly located in built-up areas, LDCs are almost invariably granted an exclusive franchise for a defined geographic zone. It would obviously be uneconomic, logistic nonsense and highly inconvenient to the public, for two or more companies to lay, own and service separate gas mains in the same streets. Thus the retailing or 'Public Distribution' of gas to small consumers is in effect 'protected' from gas-to-gas competition, but not of course from competition from oil, coal and electricity. However, as discussed in Chapter 11, even the latter is usually of a limited nature and duration. This is because once the consumer has made his investment decision to buy a cooker, space heater or some other gas-fired appliance, the cost involved in switching to another fuel is usually prohibitive, until such time as the appliance in question is no longer serviceable and has to be replaced.

Because, therefore, of the de facto monopoly that gas retailing enjoys, retail gas prices are usually controlled or regulated, directly or indirectly, at either the municipal, state or national level if only to safeguard the consumer from possible abuse. There can be, of course, other reasons of a social or political nature for the control and regulation of retail gas prices. As one would expect, much the same sort of considerations apply to the retailing of electricity and the provision of water and sewage services.

In recognition of this monopolistic position, in many countries LDCs are required by law to provide a 24 hour-day, year-round, continuous gas delivery service to firm retail customers – provided, of course, they pay their bills – under all manner of operating and climatic conditions.

Expression of the above market structure in corporate terms varies from country to country. For example, in the Netherlands, production, transmission and distribution are corporately quite separate, whereas in France transmission and distribution (with a few localised exceptions) are combined, yet again in some developing countries all activities are undertaken by one state-owned entity – there are other combinations.

Until a few years ago the United States was the classic example of this structure. Producers sold only to pipelines (transmission companies), who resold gas to utilities (LDCs), who resold gas to end-consumers. However, the supply shortages of the early 1970s and subsequent legislation led to the gas surplus of the 1980s. These events brought about a much more flexible structure and a new entrant into the chain.

Producers now sell gas not only to pipelines but also directly to large industrial consumers, electric utilities and LDCs. Pipelines also sell gas directly to industrial consumers and electric utilities as well as to LDCs. The new entrant is the Gas Marketer – see Figure 18.2. These are independent entities, or separate offshoots of transmission companies, who do not own any assets. They find producers willing to sell gas, find customers wanting gas and find appropriate transportation capacity to carry the gas they buy and sell, living on the margin they make in this process. They are not brokers as they take title to the gas and assume risks.

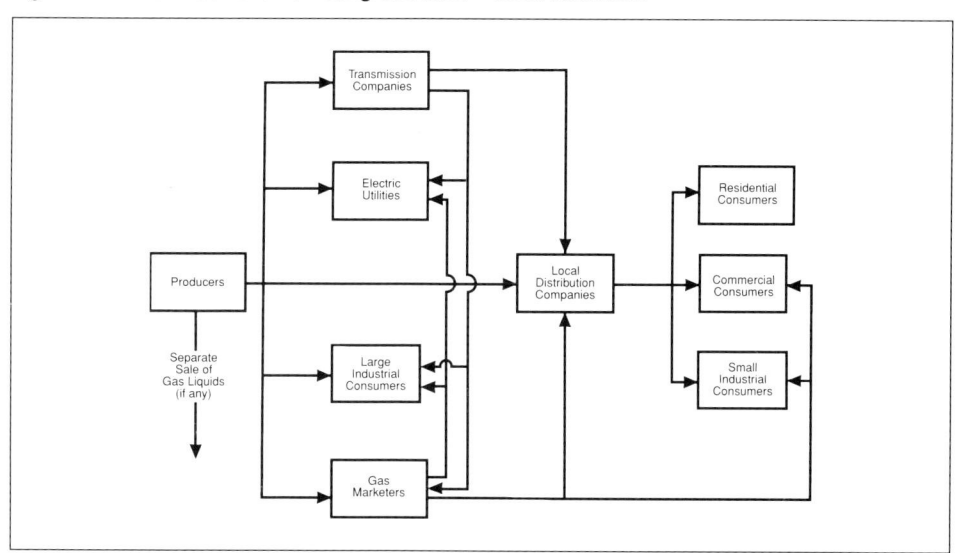

Figure 18.2 **USA: Current Marketing Structure – Multi-Activities**

Changes are also taking place in the European gas industry as market circumstances, opportunities and, in some instances, political pressures dictate. Two examples are the emergence of direct marketing by producers in the UK, and the opening up and restructuring of the East German gas industry – there are other examples with probably more changes to come.

LNG Market Structures

The structure of the LNG business is in essence much the same as the traditional structure of the pipeline gas business described above, except that it has two additional activities and, of course, shipping and trading replaces pipeline transmission – see Figure 18.3.

In the case of the LNG business, the producer sells gas at plant gate to a LIQUEFACTION COMPANY which may be privately-owned, state-owned or some combination thereof.

After liquefying the gas, the liquefaction company sells the LNG to an LNG TRADING COMPANY which usually owns or charters shipping capacity – in FOB sales the buyer provides the ships – to transport the LNG from liquefaction plant to an LNG RECEIVING TERMINAL COMPANY. The latter receives, stores and regasifies the LNG as and when required, on-selling it to either a transmission company and/or LDCs, or directly to large consumers such as an electric utility. Ownership of the terminal company varies.

Figure 18.3 **Typical LNG Base-Load Market Structure – Five Main Activities**

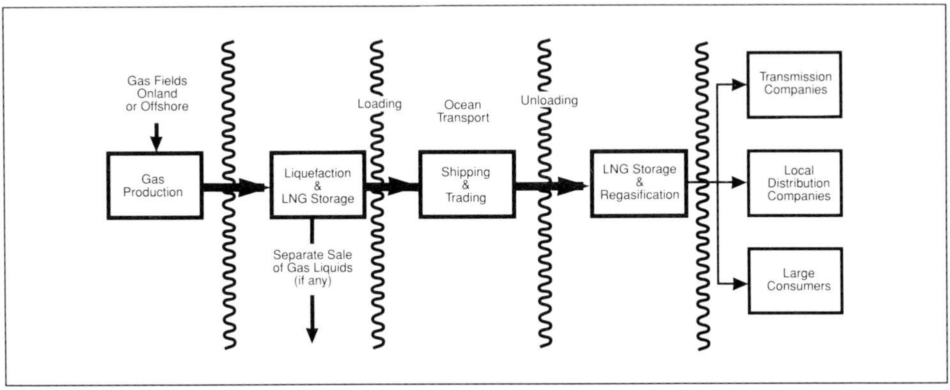

These activities may be quite separate in a corporate sense, or several of them may be the responsibility of one corporate entity. For example, with the Brunei LNG project the activities of production, liquefaction, ship ownership and trading are undertaken by separate corporate entities (although some shareholders have varying interests in all these entities) with the latter selling LNG to Japanese LDCs and electric utilities who own the LNG terminals. There are no transmission companies as such in Japan as transmission is to all intents and purposes an integral part of gas distribution due, inter alia, to the fact that all consumers are proximate to the LNG reception terminals. With the Malaysia LNG project, liquefaction, trading and ship chartering, but not ship ownership, are the responsibility of one corporate entity. As to be expected, there are other variations and combinations.

A further complexity with LNG projects is the variable point after liquefaction where title and risk to the cargo changes. This can be on an FOB (free-on-board) basis where the seller is responsible and pays for loading the LNG on to the ship. The buyer provides the ship and takes title, responsibility and risk for the LNG once it has passed the ship's loading connection.

Alternatively, it can be on a CIF (cost, insurance and freight) basis where title to the LNG passes to the buyer once the LNG is loaded or at some agreed point during the voyage. Thus the buyer is responsible for paying for the LNG once he has taken title to the LNG even if it does not arrive. However, in such an event the buyer will benefit from any insurance claimed after the risk has passed to him which is normally on loading; risk and title do not necessarily pass from seller to buyer at the same point.

The third possibility is 'delivered' or 'ex-ship' where the seller has responsibility, title and risk of the cargo until it is transferred to the buyer at the agreed delivery point. In all cases the buyer has the right to reject the cargo if it does not meet contractual specifications.

There are some variations to these three main points of sale, e.g. 'free alongside ship', 'cost and freight', etc., but they are not normally used in the LNG business. Selecting the preferred point of sale is obviously a matter for negotiation between the seller and buyer and will be conditioned by any laws and regulations of the exporting and importing countries concerned regarding trade, shipping, taxation, labour/trade unions, etc.

Irrespective of which variant is used, the underlying concept of a base-load LNG project is the dedicated supply of LNG to a dedicated buyer(s) on a long term basis using dedicated, tailor- made facilities. Unlike pipeline gas projects, for economic and technical reasons LNG projects have to be operated at very high load factors on a year-round basis. In many respects the shipping phase performs much the same sort of function as a major pipeline.

A final point: references to corporate entities in the foregoing include both incorporated entities and unincorporated joint ventures, although the latter are not very common for LNG projects – one example is the Australia North West Shelf LNG Project – but there are a number of unincorporated pipeline gas projects, notably in Europe and especially some North Sea projects.

Contracts : Some General Considerations

There is no established format or content for (wholesale) gas contracts. Each and every contract is discrete and so fashioned as to meet the needs of the individual contracting parties concerned against the backcloth of the local circumstances prevailing. This will condition not only the type of contract required, but also its degree of simplicity or complexity. Given that most wholesale term contracts have a duration of at least 10 years, sometimes 25 years or more – the United States is the main exception to this generality – contracting parties endeavour as best they can to anticipate and make provision for possible changes that could arise over contract life. In the United States, any contract of more than one year's duration is now generally regarded as a term contract.

Experience suggests that the most successful contracts are those which have been freely negotiated and accepted by both sellers and buyers with minimal constraints imposed on them by third parties, and where goodwill, mutual trust, and favourable operating experience by both parties, have enabled any subsequent revisions to contractual arrangements to be negotiated and agreed on a harmonious basis. Equally, contractual terms that unduly favour the seller or buyer have seldom stood the test of time, while any subsequent unilaterial changes that either party may attempt to introduce can cause unnecessary recriminations and, in extreme cases, possibly the collapse of the contract in question.

Finally, in those countries where government permissions and approvals are required before an otherwise freely negotiated contract can be implemented, prudent negotiators will endeavour to ensure that the appropriate authorities are kept adequately informed of those aspects of an emerging deal on which their formal agreement will eventually be sought. This is particularly important where exports are envisaged, and where governmental approvals from either the exporting and/or importing countries concerned may be required.

Main Contract Types : Pipeline Gas

The following comments are confined to wholesale term contracts between the 'seller' and 'buyer' as previously defined. For reasons already indicated, these options are unlikely to be applicable to contracts between transmission companies and LDCs and between LDCs and end-consumers.

In broad terms there are two main types of pipeline gas contracts, namely DEPLETION CONTRACTS, in which the seller dedicates all the economically recoverable reserves of a gas field or reservoir to the buyer, and SUPPLY CONTRACTS in which the seller agrees to supply individual buyer(s) with specified quantities of gas over a given period.

Typically, contracts of a depletion nature have been concluded between various southern North Sea producers of non-associated gas and British Gas in the UK, whereas Dutch gas export contracts between Nederlandse Gasunie and various buyers in Germany, Belgium, France and Italy are of a supply nature. There are, of course, many other examples of these two basic forms of contract.

The main differences between depletion and supply type contracts can be summarised as follows.

	Depletion Contract		**Supply Contract**
i)	*Involves the gas producer or producers, as one seller contracting under one contract, or as individual sellers under several similar contracts, all the economically recoverable reserves of the prescribed reservoir to one buyer, i.e. the buyer takes the reserve risk. Exceptionally, i.e. in the United States, a contract may involve several buyers.*	i)	*Usually involves one seller contracting with one buyer or several different buyers. If there are several buyers the terms and conditions are not necessarily identical for each buyer.*
ii)	*Requires a determination of reserves that are dedicated to the contract, with appropriate periodic review provisions as the reservoir is depleted.*	ii)	*Not applicable, although the buyer(s) may require evidence that adequate reserves exist to meet their and/or the totality of the supply volumes involved.*

iii) *Requires the establishment of depletion rates usually expressed in terms of average Daily Contract Quantity (DCQ) and Annual Contract Quantity (ACQ).*

iv) *Requires the establishment of annual build up rates that will apply until the agreed plateau production rate is achieved.*

v) *Requires agreement on permitted percentage swing variations of the DCQ. These are frequently stipulated for different calendar periods (i.e. weeks or months) in any year.*

vi) *Involves provisions on so-called make-up gas which set the time limits when gas not taken in any one contract year (but paid for) has to be taken. These expressions are explained later.*

vii) *The seller will usually reserve the right to develop the dedicated gas reservoir as he so wishes in accordance with good and prudent technical practice.*

viii) *The buyer may require some form of warranty that all gas dedicated to the contract is free from all adverse claims and liens.*

iii) *Not applicable. The contract will instead stipulate the total volume of gas to be supplied over contract life and usually also the maximum DCQs and/or ACQs.*

iv) *Maximum and minimum daily rates for at least the initial years of contract life are usually stipulated.*

v) *May be confined to a simple range of minimum and maximum offtake rates on a daily, monthly and/or yearly basis over contract life.*

vi) *Time limits for make-up gas may or may not be stipulated provided such gas is taken within contract life, unless contract life is subsequently extended by mutual agreement.*

vii) *Not applicable.*

viii) *Can also apply to supply contracts.*

Depletion contracts tend to be adopted where there is only one buyer, or little or no prospect of the producer being permitted or able to sell his gas to any other buyer, also where the buyer has many suppliers and a high degree of flexibility within his system. Depletion contracts were very common in the UK for many years when British Gas (and its predecessor) was then the only permitted purchaser of UK North Sea Gas. This situation no longer applies.

An advantage to the seller in this situation is that he has an assured outlet for his gas, while the buyer has the comfort that he will receive all economically recoverable reserves from the reservoir in question. Conversely, there can be some possible disadvantages, including the inability of the seller to negotiate the best possible terms from a variety of buyers, and for the buyer, the need to find markets for all the gas that will be produced. Supply contracts are more appropriate where recoverable reserves exceed an individual buyer's long term needs, and/or where other disposal options are available to the seller. This does not imply that if there is only one buyer that a depletion contract is essential, in many instances a supply contract will be more acceptable to both parties.

A third form of contract, often described as a SELLER'S OPTION CONTRACT, normally only applies where associated gas is involved. In such contracts, the buyer recognises and accepts that the prime activity and objective of the seller is to produce crude oil, with the disposal of the resultant associated gas as an important but secondary activity conditioned by the rates of crude oil production achieved or permitted.

The principal characteristic of a seller's option contract is that the seller nominates the volumes of associated gas to be sold within some agreed range. This will reflect, inter alia, the seller's expectations of crude oil recovery rates, and the gas:oil ratio of the reservoir in question which may vary as the reservoir is depleted. In most markets of significance, buyers will also have access to non-associated gas supplies and will wish to tailor their various and differing contractual arrangements to ensure that their overall supply needs are adequately satisfied and safeguarded. However, as the need to utilise high load factor associated gas, particularly where the venting or flaring of associated gas is prohibited or limited by government, may have to be accorded first priority, the accommodation of such gas may impact adversely on the offtake pattern for non-associated gas. In these circumstances, the seller will expect, not unreasonably, to receive adequate financial compensation from the buyer for providing production capacity which may only be utilised at a relatively low and highly variable load factor. All this presupposes, of course, that a market exists or can be developed in the first place. Whereas non-associated gas can, if necessary, be left in the ground until a market opportunity arises, any unrealistic conditions that may be imposed by government on the utilisation of associated gas can inhibit the regularity and maximisation of the associated oil production.

Resolution of these potential conflicts of interest and practicality, coupled always with the need for sound reservoir management, will ultimately be reflected in the respective contractual terms and conditions agreed between sellers and buyers.

Contract Type: LNG

All base-load term LNG contracts are essentially supply type contracts between one seller and one or more buyers, and as such have many features in common with pipeline gas supply contracts. An independent verification that sufficient reserves exist to support the contract may be required by the buyer.

The contract will stipulate the annual quantities of LNG that the seller will use his reasonable endeavours to supply during the build-up period of say 2 to 5 years, and within narrow limits – say up to plus or minus 10 per cent – the annual quantities to be supplied over the plateau period which typically can be 20 years. Provisions will be included for making-up any shortfalls and for the disposal of any excess quantities that may be produced.

The requirement that the seller will use all reasonable endeavours to ensure a stable, reliable and safe supply is common to many LNG contracts.

Contract Provisions: Pipeline Gas

Although many contract provisions will be common to both depletion and supply contracts, also to seller's option contracts, some provisions will be specific to one particular type of contract.

With this qualification in mind, listed below, in no special order, are examples of the variety of clauses that may be included in a pipeline gas contract. This should not be regarded as a 'check list' as it is obviously impracticable to attempt to cover all forms of contract and the many situations that can exist in different locations.

- Parties. *Who are the parties to the agreement and the date the agreement is signed by the last party to sign.*
- Definitions. *The exact meaning given by the parties of the terms and expressions used in the contract.*
- Commencement and Duration. *This may include a 'reasonable endeavours' provision to supply and take gas prior to commencement of firm contract obligations, and/or may include an option to extend the contract period at some subsequent point during the life of the contract.*
- Contract Quantities. *These may be defined in volumetric terms or as units of heat, with the latter being normally preferred. There may be a variety of further provisions to do with the contract quantities, e.g. ranges to apply daily, annually, etc., as appropriate during the build up and plateau periods of supply.*
- Transfer of Title and Risk. *Where title and risk passes from seller to buyer which may not necessarily be the same point (refer earlier comments).*
- Price. *The base contract price including indexation formula, price review procedures (if any), applicable currency, currency adjustment and taxation provisions, etc., if appropriate – see also Chapter 19.*
- Minimum Bill. *This is to ensure that the seller receives a minimum revenue to cover the costs he has incurred even if the actual gas offtake by the buyer over a defined period, or periods, is below agreed contractual offtake rates and even if Force Majeure conditions apply. Minimum bill provisions are sometimes included as an integral part of another clause.*
- Take-or-Pay. *Procedures concerning gas not taken by the buyer but paid for provided Force Majeure does not apply and the seller is able to supply. In some contracts take-or-pay provisions may form part of the minimum bill clause. Time limits and other conditions may be stipulated.*
- Make-Up Gas. *This concerns the rights of the buyer to take gas previously paid for but not taken. This can be an integral part of the minimum bill or take-or-pay clauses. The period for which this right can be carried over is important and a limit of a certain period will usually be negotiated and expressly provided.*

- Nominations. *Procedures whereby sellers and buyers agree in advance on the quantities to be supplied during a given period, e.g. day, week, month, etc. These procedures help to ensure, inter alia, that the seller can meet the buyer's offtake requirements within contractual limits, and enable the seller to schedule his essential maintenance periods. Provisions for seasonal fluctuations in offtake over the year are often a crucial issue.*
- Quality. *This will normally specify the minimum and maximum calorific values of the gas to be supplied, also its composition in terms of the hydrocarbons, non-hydrocarbons, including impurities, that is acceptable to the buyer.*
- Delivery. *The point or points where the seller agrees to deliver gas to the buyer, including, or as a separate clause, the pressure at which the gas is to be supplied.*
- Taxes. *Responsibility for the payment of any taxes and similar charges levied by government, or governments in the case of international sales, over and above those normally applicable, e.g. corporation tax, to the seller's and buyer's usual business activities.*
- Permissions and Approvals. *Seller and buyer use their best endeavours to obtain and maintain all necessary permissions and approvals to enable them to perform their contractual obligations.*
- Facilities. *Responsibility for the provision of facilities for the delivery and receipt of supplies.*
- Measurement. *Metering/measurement arrangements and applicable standards, including procedures for the periodic validation of measuring equipment.*
- Gas Liquids. *Extraction rights for and/or the ownership of gas liquids (if applicable). In most conventional pipeline gas sales contracts this will not arise as the seller will have already removed entrained liquids to the level necessary to meet the buyer's requirement for a gas of marketable quality.*
- Invoicing and Payment. *Procedures as to how, when, to whom and where payment is to be made, the frequency of presentation and payment of invoices, and how disputed invoices and overdue payments shall be handled.*
- Renegotiation. *Provisions, which may embrace pricing, quantities and/or other matters as may be appropriate to the contract. More often than not such provisions, if any, will form part of the clauses in question.*
- Applicable Law and Language. *The law and language under which the agreement is construed and will be governed.*
- Settlement of Disputes. *The resolution of disputes by experts and arbitration procedures in the event that the parties to the contract cannot resolve any dispute, controversy or claim between themselves amicably.*
- Termination and Notice Period. *Self explanatory, but may not be included in a fixed-term contract.*

- Force Majeure. *Provisions which excuse the failure of a party to fulfil any of his obligations, other than the obligation to pay money, to the extent these are caused by circumstances beyond his reasonable control.*
- Confidentiality. *Procedures which govern the disclosure of information by either the seller or buyer to third parties.*
- Default. *Procedures, including arbitration arrangements, in the event of either party failing to meet his contractual obligations.*
- Assignment. *Procedures in the event of either party wishing to assign his contractual obligations to some other party.*
- Exchange of Information. *Co-operation on planning matters and the exchange of information necessary to enable both parties to fulfil the terms and obligations of the contract.*
- Notices. *Where, how and to whom notices are to be served.*

At the risk of repetition, the above list is illustrative and not exhaustive, and some of the clauses listed above may not be applicable to certain types of contract. Obviously, for depletion contracts a determination of reserves will also be required, and procedures may need to be established in the event that the producer is able to offer additional gas which the buyer may wish to purchase.

Prospective buyers will no doubt also require evidence of the seller's right to produce gas and from where, and to be assured that all necessary governmental licenses and consents to produce the gas have been granted. These and other matters may or may not form part of the formal contract, but if not they will most probably arise and be discussed during contract negotiations and/or be covered in side letters or schedules to the contract.

Contract Provisions: LNG

Most of the provisions listed above for pipeline gas contracts also apply to LNG contracts, all of which are supply type contracts. However, there are some other provisions which are unique to LNG contracts. They include:

- Procurement. *The party responsible for the procurement of ships.*
- Transportation. *Stipulates the maximum dimensions and cargo capacities of the ships that will be used, and their timing, manner and responsibility for arrival, berthing, loading, unloading and departure and for any delays that may be incurred.*
- Delivery. *Establishing each year an annual delivery programme. This is similar to but not entirely the same as the nomination procedures for pipeline gas contracts if only because pipeline gas is a continuous, if variable, supply, whereas LNG shipments, by definition, are intermittent supplies.*
- Safety. *Particularly as regards the crewing, operation and maintenance of LNG ships.*

In cases where the seller in fact comprises several sellers in an unincorporated joint venture, each seller will probably conclude its own contract with the buyer(s) concerned for its share of the total contract quantity to be supplied. Each seller's contract would contain terms which are substantially the same, except in relation to the quantities to be supplied which will vary according to the seller's equity interest in the joint venture.

Some Concluding Observations

Gas contracting is a very complicated and time-consuming affair. While some provisions can be quickly agreed, others may take months or even years of negotiation before they can be expressed in final contract form. Pricing is almost invariably the most difficult and last issue to be resolved.

Where goodwill, mutual trust and firm intent exists between seller and buyer, commitment to and construction of the project can sometimes proceed in advance of a final, definitive contract by concluding a 'Letter of Intent' or some similar preliminary agreement. Even the contentious issue of price can be handled by agreeing that, for example, it shall not be less than X or more than Y before it is finally and precisely quantified. However, it has to be stressed that commitment in advance of a signed sale and purchase contract is very much the exception and would only be adopted in special circumstances.

The focus of this chapter has been on wholesale term contracts between buyers and sellers for pipeline gas, indicating certain special features which apply to LNG contracts. In practice, a whole host of other contracts will need to be negotiated and concluded with landowners, contractors, banks, insurance houses, trade unions, local authorities, equipment manufacturers, shipowners, shipbuilders, etc., as well as with those entities which will be responsible for designing, supervising construction, operating and maintaining the project's facilities.

Contract drafting is very much the outcome of a team effort in that the drafters need to draw upon and take account of all the technical, commercial, financial, legal, fiscal and other expertise available to them. Even then it is more an art than a science in that it is virtually impossible to foresee and provide for every eventuality that might arise over the future life of the contract.

Chapter 19

NATURAL GAS PRICING

Because pricing is the ultimate expression of the gas business in monetary terms, the author considered it to be the appropriate topic with which to conclude this book. It will always be subject to debate, critical analysis, frequent appraisal and new concepts as gasmen continue to seek that elusive optimum formula which will stand the test of time through virtually any changes in market circumstances. It is a sensitive area with confidentiality often requested by seller and/or buyer or, in some instances, imposed by government. By contrast, in those countries where pricing arrangements are matters of public record, their transparency does not necessarily ensure that they are more soundly based, endurable, or 'fairer' to all parties concerned than those that are not. In the final analysis, pricing is all about market perceptions, supply and demand, competitive influences, balancing risks and rewards and, above all else, negotiating skills – it is hugely time-consuming.

Some General Concepts and Principles
The essence of a successful trade contract is that goods, in this instance natural gas, are supplied to the buyer under conditions, including pricing arrangements, which satisfy and are fair to both buyer and seller. These conditions become more important as the duration of the contract increases, since they must not only satisfy the longer term aspirations of both buyer and seller, but must also be responsive to changes in the general business environment which affect both the usefulness of the goods and the participants' aspirations.

Natural gas contracts differ markedly from oil contracts in a number of respects. Some of the more fundamental differences are:

i) *Unlike crude oil, there is no generally recognised international 'marker' price or pricing system for traded gas. Pricing of natural gas has historically been negotiated for individual contracts, although with an increasing measure of commonality in some regional markets. Moreover, gas prices have generally been more responsive to local conditions, particularly for gas which is consumed within the country in which it is produced. Over 85 per cent of the gas consumed in the world today is locally produced.*

ii) *Although oil and gas exploration and production facilities may involve similar levels of investment, transportation and distribution systems for gas, particularly for supplying multiple small scale users in the residential and commercial sectors, require significant additional investment. In such circumstances the buyer needs adequate assurance of security of supply to utilise his large investment efficiently. At the same time, because the seller usually has few, if any, alternative outlets available to him, he will in turn need assurance of security of offtake from the buyer.*

iii) *Because the costs of gas transportation per unit of energy are much higher than for oil, and because gas supply systems are inherently inflexible, there is a high degree of interdependence between seller and buyer and a need for long term commitment which binds the buyer and seller of gas into a lasting commercial relationship.*

Important though these and other differences are, perhaps the most significant difference between the oil and gas business is that, for oil production, investment can be made stepwise and usually precedes the securement of sales contracts, whereas for gas long term contracts are almost invariably made between prospective seller and buyer before development investment is undertaken.

Another important characteristic of the gas business is that the balance of technical, financial and political risks inherent in any large gas venture tends to weigh more heavily on the seller (producer) than on the buyer. There are, of course, risks on the buyer's side as well, but they are usually of a more manageable and containable nature. These risks need to be recognised by sellers and buyers in their contract conditions and reflected through the risk/reward concept, the single most important expression of which is price.

Gas contracts are both complex and specific and take a variety of forms. They embody provisions covering a large number of elements and operating factors: typical provisions and contract types are described in Chapter 18.

A gas project usually involves a number of separate entities at various stages of the chain from production source to end-consumer. Contracts define the role of each

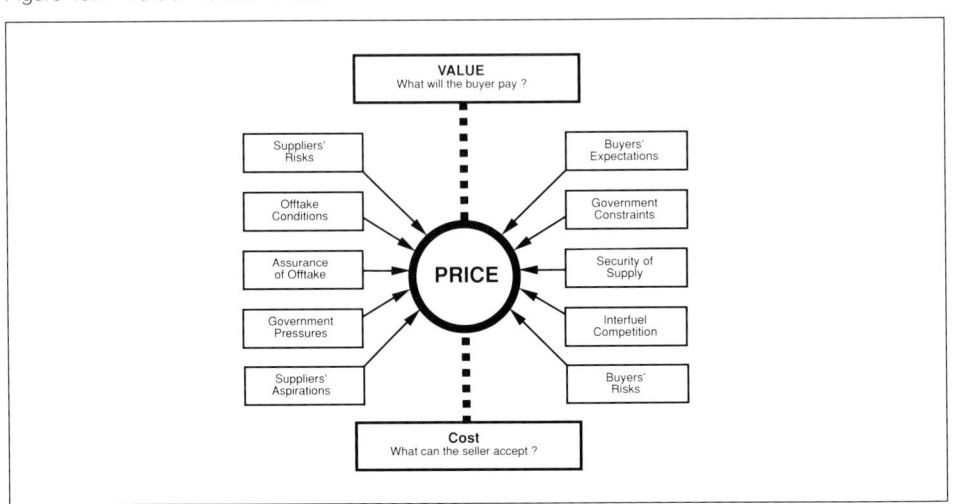

Figure 19.1 **Value – Cost – Price**

entity and, by recognising their respective obligations and risks, set out mutually agreeable rewards.

The principal exception today to some of the foregoing generalisations is the pipeline gas business in the United States, the world's oldest and largest gas market after the Soviet Union, where special circumstances exist as discussed later.

Price Components

Three important components that usually appear in the price clause of a pipeline gas contract are described below. In this context these comments are directed principally at the pricing basis between the seller (producer) on the one hand, and the buyer (transmission company) on the other hand. In other words they relate to what is usually termed wholesale prices and as such may not necessarily apply to retail or end-consumer prices.

THE COMMODITY CHARGE, or the base price per unit of gas expressed as a unit of currency for a specified unit of energy/heat at some agreed point in time, e.g. US dollars per million Btu, UK pence per therm (100,000 Btu), Dutch florins per normal cubic metre, etc. If a volumetric unit is used then its heat content will be specified.

INDEXATION PROVISIONS, the method by which the commodity charge may be varied up or down at agreed time intervals to reflect changes in the energy market and/or the business environment.

Indexation provisions can be very simple or quite complex. A simple form of indexation would be where the base price (commodity charge) is indexed with changes in one index, say the published price of fuel oil or crude oil or some official index such as a Wholesale (or Retail) Price Index. For example:

$$\text{Prevailing price} = Po \times \frac{WPI}{WPIo}$$

Where:

Po = *base price of the gas*

WPI = *Wholesale Price Index over the relevant time period*

WPIo = *base Wholesale Price Index*

Simple forms of indexation of this type were quite common until the early 1970s, but thereafter as energy prices and inflation rates became much more volatile, complex provisions were introduced to better reflect a less predictable, less stable business environment, and also a more varied use of gas itself. An example of this is:

$$\text{Prevailing price} = Po \times \left\{ 0.45 \frac{GO}{GOo} + 0.40 \frac{FO}{FOo} + 0.15 \frac{WPI}{WPIo} \right\}$$

Where:

Po = *base price of gas*

GO = *gas oil price over the relevant time period*

GOo = *base gas oil price*

FO = *fuel oil price over the relevant time period*

FOo = *base fuel oil price*

WPI = *Wholesale Price Index over the relevant time period*

WPIo = *base Wholesale Price Index*

0.45, 0.40 and 0.15 = respective proportions of the indices applied

These examples are illustrative only with the last example being representative of a wholesale pipeline gas contract where the main competitive fuels are considered to be gas oil and fuel oil. None of these examples should be regarded as being models for using in new contracts.

No indexation system is perfect. During the life of a long term contract it may be necessary from time to time for the seller and buyer to renegotiate changes so that the indices are more reflective of the current energy climate and/or of the balance of competitive fuels in an evolving market.

However, such changes do not resolve one continuing imperfection. This is the inevitable delays involved before reliable (published) data are available to trigger the indexation provision(s). Time lags of at least a month, often several months, are commonplace. This can result in the applicable gas price being increased while actual competitive fuel prices are falling, and vice versa – see Figure 19.2.

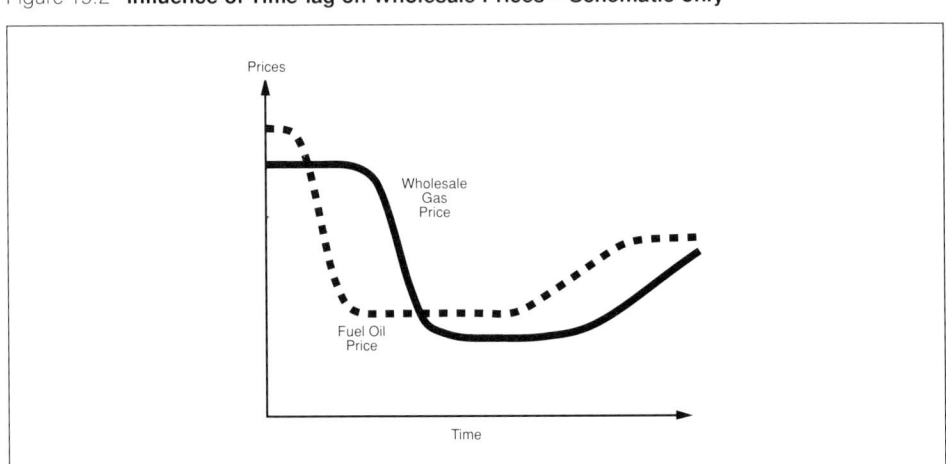

Figure 19.2 **Influence of Time-lag on Wholesale Prices – Schematic only**

Nevertheless, these discrepancies are usually manageable and in any event tend to balance out over time.

The third main component is the CAPACITY CHARGE. This reflects all or part of the cost to the seller of providing production and/or transport (pipeline) capacity which may be utilised at varying hourly, daily, monthly and/or yearly rates (or load factors) by the buyer under the terms of the contract.

Load factors can be expressed in time periods or as percentages. For example:

$$\text{Load factor in days} = \frac{\text{annual throughput}}{\text{throughput on the peak day}}$$

$$\text{Load factor in hours} = \frac{\text{annual throughput}}{\text{throughput in the peak hour}}$$

$$\text{Load factor as a percentage} = \frac{\text{annual throughput}}{\text{peak throughput}} \times 100$$

If both throughputs are expressed 'per day', then the expression gives the daily load factor, and if 'per hour' the hourly load factor. Most English-speaking countries express load factors as percentages, whereas most non-English-speaking countries prefer hours or days, but practices do vary.

The expression:
$$\frac{\text{average throughput}}{\text{maximum capacity of the system}} \times 100$$

gives the 'utilisation factor' not the load factor.

Other components, particularly for international contracts, may need to be included to provide for currency fluctuations, or to 'smooth out' large and abrupt changes in the indices used.

None of these three main components need necessarily appear in a particular contract, but all of them should be very carefully considered in approaching the basic principles of pricing which centre around:

i) *A minimum price the seller is able to accept, within the context of other contractual conditions, and which rewards his risks in order to proceed with the investment required.*

ii) *A maximum price the buyer is willing and able to pay, given the alternatives, if any, that may be open to him.*

Where a reasonable price range is defined in this way, negotiations will normally settle at a price somewhere between the two extremes of the range that is acceptable to both parties. If the range is narrow, however, price negotiations, and particularly indexation to reflect future uncertainty, become crucial to a stable relationship between willing buyer and willing seller.

If, as can sometimes occur, the range is negative, i.e. the seller's minimum price is above the buyer's maximum price, one or both parties may have to revise their aspirations. Experience suggests that, without a definitive and equitable settlement, trouble is merely postponed with the possibility, under adverse circumstances, that the contract may eventually break down to the detriment of both parties.

A brief word at this point on retail (tariff) prices to small consumers. These usually comprise two components, a commodity charge and a standing charge, the latter charge being a fixed monthly, quarterly or annual amount of money the consumer has to pay irrespective of how little or how much gas he uses. In many countries, either or both these charges are regulated by a local or national official body whose approval is required before any changes can be made by the gas (distribution) utility concerned. In some instances, price regulation embraces large consumers as well.

LNG Pricing

The main thrust of the foregoing has been on the wholesale pricing of pipeline gas, although many of the principles and concepts discussed apply equally to the wholesale pricing of LNG. One of the main differences is that with LNG pricing the capacity charge, as described above, is not applicable. Instead, LNG term contracts stipulate within a narrow range the annual quantities to be supplied/delivered, and the regularity of such deliveries over the year, once the plateau rate of supply is reached. By definition, base-load LNG projects have to be operated at a high, year-round rate (load factor) for economic and logistical reasons.

The pricing formula – complemented by separate clauses on take-or-pay, minimum bill and the like (see Chapter 18) – is usually limited to a base price (commodity charge) and indexation, with possibly some review or adjustment provision. In the following example both the base price and indexation are related to crude oil prices. For some years, crude oil prices have been the preferred pricing basis for all operational base-load LNG projects serving Pacific Basin markets, and for most other LNG projects elsewhere.

$$\text{Prevailing price} = Po \times \frac{C(n-1) + A}{Co}$$

Where:

Po = *base price of LNG*

C(n-1) = *the average weighted Government Selling Price of a specified cocktail of crude oils imported by the country in question on the last day of a specified 'n-1'.*

Co = *a specified crude oil Government Selling Price.*

A = *an adjustment factor, with specified monetary limits, applicable to LNG sold in month 'n' as may be negotiated from time to time between seller and buyer to reflect market conditions.*

As with previous examples this is illustrative only. In practice, each and every pricing formula, whether it be for LNG or pipeline gas, will be different and reflect the outcome of a negotiation between seller and buyer and their perception of the market at some point in time.

Pricing Options

There are three principal pricing options used in term contracts, although there are other variations in practice within this general framework. More recently, a fourth form of pricing – spot prices – has evolved in North America which, as the name implies, only applies to gas sold on a very short term, once-off basis.

The FIXED PRICE option is where the commodity charge is the price fixed for the duration of the contract period, with indexation and the capacity charge playing little or no part. This form of pricing was used in many contracts, with variations, up until the early 1970s, longer in some cases. Examples include the initial contract arrangements for Alaskan and Brunei LNG supplied to Japan, UK southern North Sea pipeline gas supplied to British Gas (in this instance with a limited element of indexation with inflation), and regulated producer prices in the United States which also had incremental adjustments as determined by the regulatory authority – there are other examples. In all the cases quoted here, circumstances have changed since the initial arrangements were established.

The principal disadvantage of fixed pricing is that it lacks flexibility in the face of change. Thus while it may be satisfactory at times when inflation is low and competing energy prices are relatively stable or flat, it provides little or no incentive for producers to explore for and develop new gas when these external conditions no longer apply. As already mentioned, this was the case following the oil price shocks of the 1970s. Many fixed price contract arrangements were abandoned around that time and replaced with pricing arrangements which recognised the need for greater flexibility.

Thus while a fixed price system may embody contract re-openers, or the chance to renegotiate prices and other terms, without formal linkages between the gas price and external market indicators, such renegotiations can be as time consuming and adversarial as the original, and at best restart the cycle.

Moreover, fixed price contracts tend to imply that gas is priced lower, sometimes substantially lower, than competitive fuels other than those of a primitive nature, e.g. wood, or of a less convenient nature, e.g. coal. As a consequence, the demand for gas in countries where the full range of alternative energies are available can be artificially stimulated, with increased penetration into perhaps less attractive market sectors, and, to a degree, less efficient usage. In some countries, this distortion subsequently led to the promotion of higher-cost imported gas supplies as indigenous producers became increasingly reluctant to embark upon high-risk exploration, with inadequate rewards, for new gas reserves.

The problem is a little more complex in developing countries where gas may initially replace very low cost fuels. Nevertheless, even in such cases, the real alternative on grounds of overall efficiency and the longer term economic well-being of the country, is more likely to be the higher quality fuels such as oil and electricity than more primitive forms of energy. Much will depend here on consuming government policies and aspirations, and the desire to attract outside investment, particularly in the energy and related fields.

To sum up, fixed pricing has little to commend it and many inherent disadvantages in an energy environment subject to large and sudden changes.

Under the second option, COST-PLUS, the buyer pays a price consisting of the seller's costs, together with an agreed margin. This option has been used in a number of developing countries, and usually lies at the bottom of the pricing range discussed earlier. In some instances it may be the only available and acceptable option, always provided that the rewards in relation to the risk money involved are acceptable.

From the buyer's viewpoint, this option has the apparent advantage of making gas available at what is, or appears to be, a 'low' cost. However, it raises a number of problems:

- *Uncertainty, in the form of inflation, exchange rate fluctuations, changes in the world energy environment, technical factors such as difficult field conditions, and so on. Moreover, it can substantially raise the price above the buyer's original expectations, particularly over the course of a long term field development programme.*

- *It poses the question of what is an 'acceptable' return (or 'plus') to the seller on his investment?*

- *How is risk recognised and rewarded for current and future exploration and production activities?*

- *There is little incentive for the seller to be cost efficient if he knows that all the costs he incurs will be recovered.*

- *Lastly, the return to the producer may not match rewards available for similar opportunities elsewhere, with the result that it could put future supplies in jeopardy.*

By definition, indexation directly to changes in the external environment plays no part in this pricing system, and, as indicated, the commodity charge being cost-dependent, is essentially variable, while any reflection of the capacity charge becomes unnecessarily complex or is abandoned.

Cost-plus is generally not a preferred option, since it fails to recognise the effect of market forces, and not infrequently breaks down at some point during the life of the contract.

With the MARKET VALUE option the price of gas is related in some way to its perceived or actual value to final users. Because end-use market values tend to change over time, sometimes quite suddenly, indexation is likely to play a major role in this type of pricing structure. Sellers and buyers have increasingly moved to this type of contract since the early 1970s for reasons indicated earlier.

The key question is, of course, what the market value of the gas is assessed to be. There are a number of ways of establishing this:

i) *Fuel substitution – if the gas is substituted for, say, an oil product, there is a 'cost saving' for that fuel, which may be measured in terms of international oil prices, and/or the value of the alternative use for that oil saved.*

ii) *Gas replacement – in a situation of multiple gas sources or excess gas availability, the value of one gas source can be measured against the cost of alternative supplies, whether indigenous or imported via pipeline or as LNG.*

iii) *End product value – where a product is produced using gas as a feedstock or a fuel, its value can be measured either against an alternative way of making that product (i or ii above), or against the cost of importing that product, or in terms of not having the product available at all.*

For example, if it is proposed that gas be used in a fertiliser plant, the options to be compared to value the proposal are:

- *gas as feedstock;*
- *other fuel as feedstock – at a cost of indigenous production or importing;*
- *imported fertiliser – at what are probably likely to be international trading prices;*
- *no fertiliser – at a measurable cost in terms of agricultural productivity.*

In the same way, the value of gas for, say, residential purposes may be measured in terms of fuel substitution e.g. LPG, kerosene, domestic heating oil, etc., or quantified in a less formal way in terms of social benefits of home heating and convenience for cooking and water heating.

The measurement of costs is not only a matter of fuel costs, but involves total resource costs. These include infrastructure investment, capital costs for end-users, operating costs, storage costs for wholesale buyers and consumers, as well as the original production costs and the price paid for delivered fuel. In some circumstances it may also be appropriate to consider the spin-off effects for local employment, energy security and so on, which in most instances are less directly measurable.

Market value related pricing systems have the major benefit of placing each specific pricing decision into a much wider context which embodies energy availability and pricing in both a national and international perspective, and, as already indicated, should take account of national energy policy and aims. If the base price and indexation system are set up appropriately, gas prices should move in ways which allow gas to fulfil its allotted role in the national energy structure, while continuing to provide existing and prospective producers with sufficient incentives to stimulate further exploration and the development of supplies.

In practice, these many considerations have to be condensed to manageable terms. For example, if gas is to be sold predominantly for industrial purposes, fuel oil (low or high sulphur content) may be considered to be the main 'market value' for gas. Similarly, gas oil, i.e. domestic heating oil, for the residential market. In many markets, however, where gas is used for a variety of applications, such simple relationships may not be appropriate and a more complex mix of market values has to be established. A further factor which may apply, especially where environmental concerns are important, is how to accord gas a premium value that recognises its clean burning characteristics.

In sum, the determination of market value can be as simple or as complex as the seller and buyer wish it to be, but the end result must be manageable, comprehensible and relatively easy to apply in pricing terms.

The Spot Market and Spot Prices

The fourth form of pricing, SPOT PRICES, is not so much an option, but more an outcome of a market situation. Apart from a few occasional deals, which are usually short term increments to existing term contracts, the spot gas business of any magnitude is currently confined to the United States market.

In that market current levels of demand for gas are still way below the peaks of the early 1970s, with the result that there exists a greater production and deliverability capacity than the market requires, supported by a huge storage capacity to cope with high seasonal offtakes. This, together with progressive deregulation and related enabling legislation, has created severe gas-to-gas competition as producers and transmission companies seek to utilise their available capacity to the maximum extent possible. Term contracts still exist and continue to be concluded, but a large proportion of gas is sold on a very short term basis – typically as little as 30 days – at the going spot price. Market value in the context described above, indexation provisions and the like do not apply; producers and transporters principal objective is to utilise their available capacity to generate revenue which hopefully more than covers their sunk investment and operating costs. Although there are some superficial similarities, the spot business should not be confused with the cost-plus pricing option.

Spot business, and its corollary spot prices, can only exist on any scale and for any length of time in a mature market where the buyer(s) feel sufficiently confident that their supply needs can be satisfied adequately and immediately, virtually on a day-to-day basis. From about 5 per cent of total gas consumption in 1983, the spot market grew to an estimated 75 per cent by 1988, and while it has declined somewhat since then, it still comprised more than half of total consumption by the early 1990s. Interestingly, reported and posted spot prices are used increasingly for indexing the pricing provisions in new term contracts.

Another feature of the spot business was the establishment in April 1990 of a natural gas futures trading market. In simple forward hedges, buyers and sellers reduce their vulnerability to unforeseen changes in spot market prices by taking counter- positions in the futures market. More complex plays include intermarket hedges (e.g. gas to oil), and intramarket hedges, one time frame versus another.

It remains to be seen if the spot market and spot prices will survive as and when production and deliverability capacity closely match or become insufficient to meet demand.

Summary

Price is the ultimate reward for the risks taken and for the goods (natural gas) and services provided. Experience suggests that market value related systems usually provide the most realistic and consistent mechanism for achieving a stable, fair and continuing relationship between buyer and seller, particularly in a world of increasing change. Nevertheless, it is acknowledged that situations and circumstances can vary quite considerably, on occasion other pricing systems may be more appropriate in order to get the business up and running, or because they are imposed upon the seller and buyer by government, or because they are the outcome of special local market characteristics as in the United States. But whatever pricing system is adopted, it must obviously be fair and remunerative for both contracting parties otherwise both will suffer if either party is put out of business.

NATURAL GAS AND OTHER ENERGY EQUIVALENTS

The following are quick-reference equivalents. All figures are APPROXIMATE VALUES only for use where precision is not required. They are based on:

(i) *for natural gas:*
    $1,000$ Btu/ft^3 = $9,500$ kcal/m^3 (Groningen gas $8,400$ kcal/m^3)

(ii) *for LPG:*
    an assumed 50/50 propane/butane mixture with (r) or (p) indicating that the LPG is either refrigerated or pressurised.

(iii) *calorific values, MMBtu (gross):*

per tonne	– LNG 51.8; LPG 47.3; oil 42.3; coal 27.3
per barrel	– LNG 3.8; LPG (r) 4.45; LPG (p) 4.1; oil 5.8
per cubic metre	– LNG 23.8; LPG (r) 28; LPG (p) 25.8

Table 1 **Natural Gas: Cubic Metre Equivalents**

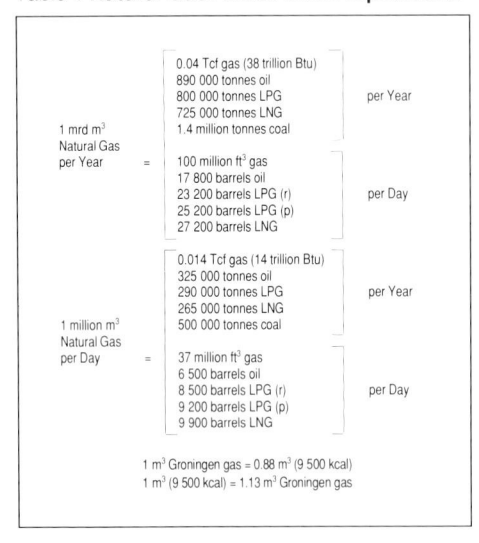

1 mrd m^3 Natural Gas per Year =
- 0.04 Tcf gas (38 trillion Btu)
- 890 000 tonnes oil
- 800 000 tonnes LPG
- 725 000 tonnes LNG
- 1.4 million tonnes coal

per Year

- 100 million ft^3 gas
- 17 800 barrels oil
- 23 200 barrels LPG (r)
- 25 200 barrels LPG (p)
- 27 200 barrels LNG

per Day

1 million m^3 Natural Gas per Day =
- 0.014 Tcf gas (14 trillion Btu)
- 325 000 tonnes oil
- 290 000 tonnes LPG
- 265 000 tonnes LNG
- 500 000 tonnes coal

per Year

- 37 million ft^3 gas
- 6 500 barrels oil
- 8 500 barrels LPG (r)
- 9 200 barrels LPG (p)
- 9 900 barrels LNG

per Day

1 m^3 Groningen gas = 0.88 m^3 (9 500 kcal)
1 m^3 (9 500 kcal) = 1.13 m^3 Groningen gas

Table 2 **Natural Gas: Cubic Foot Equivalents**

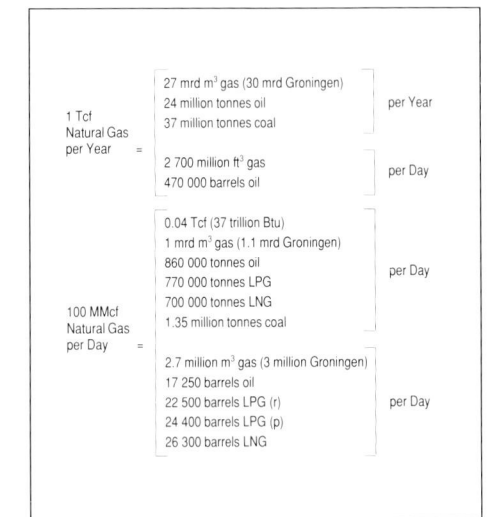

1 Tcf Natural Gas per Year =
- 27 mrd m^3 gas (30 mrd Groningen)
- 24 million tonnes oil
- 37 million tonnes coal

per Year

- 2 700 million ft^3 gas
- 470 000 barrels oil

per Day

100 MMcf Natural Gas per Day =
- 0.04 Tcf (37 trillion Btu)
- 1 mrd m^3 gas (1.1 mrd Groningen)
- 860 000 tonnes oil
- 770 000 tonnes LPG
- 700 000 tonnes LNG
- 1.35 million tonnes coal

per Day

- 2.7 million m^3 gas (3 million Groningen)
- 17 250 barrels oil
- 22 500 barrels LPG (r)
- 24 400 barrels LPG (p)
- 26 300 barrels LNG

per Day

Table 3 LNG: Volumetric Equivalents

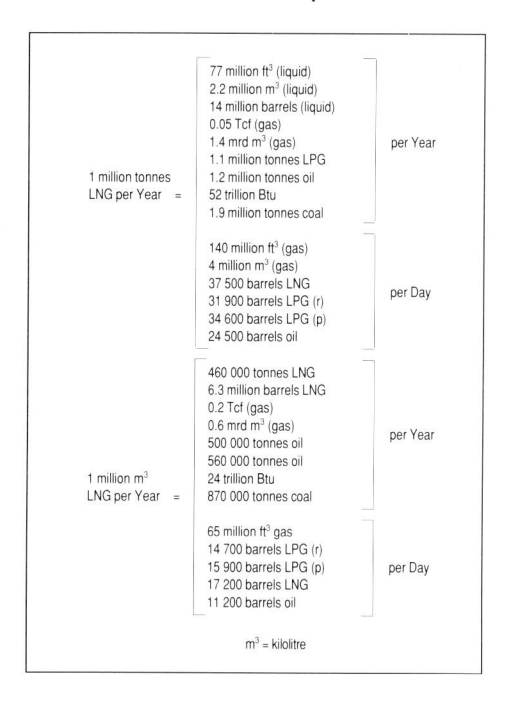

1 million tonnes LNG per Year =

- 77 million ft³ (liquid)
- 2.2 million m³ (liquid)
- 14 million barrels (liquid)
- 0.05 Tcf (gas)
- 1.4 mrd m³ (gas)
- 1.1 million tonnes LPG
- 1.2 million tonnes oil
- 52 trillion Btu
- 1.9 million tonnes coal

per Year

- 140 million ft³ (gas)
- 4 million m³ (gas)
- 37 500 barrels LNG
- 31 900 barrels LPG (r)
- 34 600 barrels LPG (p)
- 24 500 barrels oil

per Day

1 million m³ LNG per Year =

- 460 000 tonnes LNG
- 6.3 million barrels LNG
- 0.2 Tcf (gas)
- 0.6 mrd m³ (gas)
- 500 000 tonnes oil
- 560 000 tonnes oil
- 24 trillion Btu
- 870 000 tonnes coal

per Year

- 65 million ft³ gas
- 14 700 barrels LPG (r)
- 15 900 barrels LPG (p)
- 17 200 barrels LNG
- 11 200 barrels oil

per Day

m³ = kilolitre

Table 4 LPG & Ethane: Weight, Volume, Heat Conversions

BARRELS PER TONNE

	Ethane	Propane	n-Butane	C_3/C_4 mix
Pressurised	17.6	12.4	10.8	11.6
Refrigerated	11.5	10.8	10.4	10.6

CUBIC METRES PER TONNE

	Ethane	Propane	n-Butane	C_3/C_4 mix
Pressurised	2.80	1.97	1.71	1.84
Refrigerated	1.83	1.72	1.66	1.69

10^6 Btu PER BARREL

	Ethane	Propane	n-Butane	C_3/C_4 mix
Pressurised	2.79	3.85	4.35	4.10
Refrigerated	4.27	4.41	4.49	4.45

10^6 Btu PER CUBIC METRE

	Ethane	Propane	n-Butane	C_3/C_4 mix
Pressurised	17.6	24.2	27.4	25.8
Refrigerated	26.9	27.7	28.3	28.0

10^6 Btu PER TONNE

	Ethane	Propane	n-Butane	C_3/C_4 mix
Pressurised Refrigerated	49.2	47.7	46.9	47.3

1 BARREL/DAY = TONNES PER ANNUM

	Ethane	Propane	n-Butane	C_3/C_4 mix
Pressurised	20.7	29.4	33.8	31.6
Refrigerated	31.7	33.8	35.0	34.4

Table 5 Natural Gas: Inter-fuel Price Equivalents (US Currency)

ONE CENT per MMBtu =

crude oil	5.8
fuel oil	6.4
naphtha	5.2
LPG (r)	4.5
LPG (p)	4.1
LNG	3.8

CENTS per BARREL

ONE CENT per MMBtu =

crude oil	0.42
fuel oil	0.40
naphtha	0.45
LPG	0.47
LNG	0.52

DOLLARS per TONNE

ONE DOLLAR per BARREL =

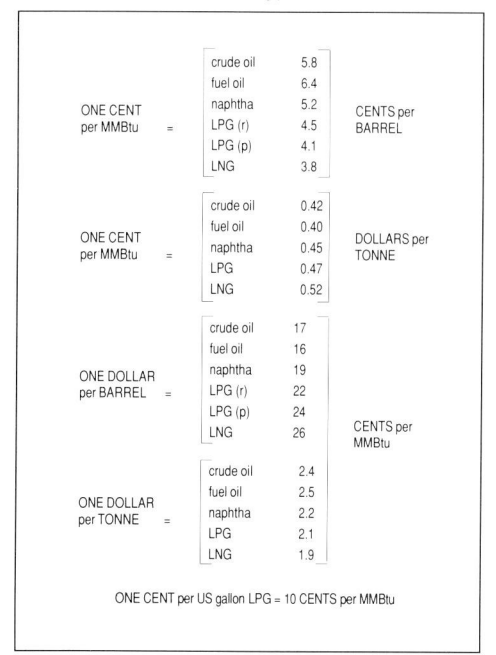

crude oil	17
fuel oil	16
naphtha	19
LPG (r)	22
LPG (p)	24
LNG	26

CENTS per MMBtu

ONE DOLLAR per TONNE =

crude oil	2.4
fuel oil	2.5
naphtha	2.2
LPG	2.1
LNG	1.9

ONE CENT per US gallon LPG = 10 CENTS per MMBtu

Table 6 Oil And Coal Equivalents

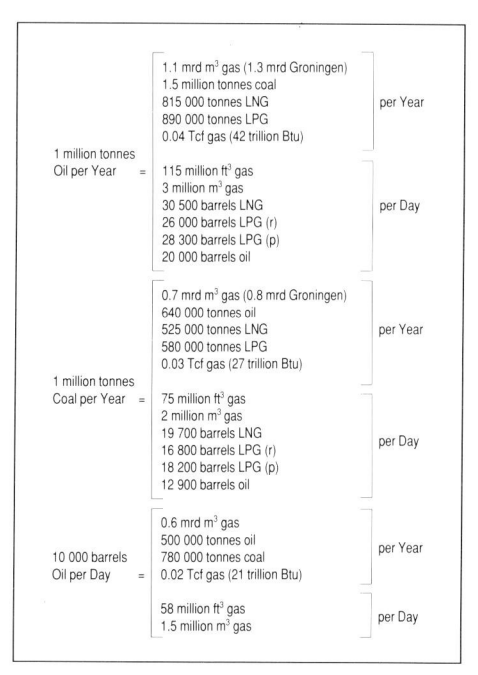

1 million tonnes Oil per Year =

- 1.1 mrd m³ gas (1.3 mrd Groningen)
- 1.5 million tonnes coal
- 815 000 tonnes LNG
- 890 000 tonnes LPG
- 0.04 Tcf gas (42 trillion Btu)

per Year

- 115 million ft³ gas
- 3 million m³ gas
- 30 500 barrels LNG
- 26 000 barrels LPG (r)
- 28 300 barrels LPG (p)
- 20 000 barrels oil

per Day

1 million tonnes Coal per Year =

- 0.7 mrd m³ gas (0.8 mrd Groningen)
- 640 000 tonnes oil
- 525 000 tonnes LNG
- 580 000 tonnes LPG
- 0.03 Tcf gas (27 trillion Btu)

per Year

- 75 million ft³ gas
- 2 million m³ gas
- 19 700 barrels LNG
- 16 800 barrels LPG (r)
- 18 200 barrels LPG (p)
- 12 900 barrels oil

per Day

10 000 barrels Oil per Day =

- 0.6 mrd m³ gas
- 500 000 tonnes oil
- 780 000 tonnes coal
- 0.02 Tcf gas (21 trillion Btu)

per Year

- 58 million ft³ gas
- 1.5 million m³ gas

per Day

Page 241

SYMBOLS AND ABBREVIATIONS

Not all of the following necessarily conform with the International System of Units (SI) as recommended by the International Gas Union. Nevertheless, those that do not conform with SI are still widely used by the gas industry as a convenient, if inconsistent, shorthand for everyday purposes.

Correct use of capital and small letters is essential. For example, mm is the SI unit for millimetre and should not be confused with MM which is the convenient symbol for million. Points or full stops are not used. Abbreviations have the same form for singular or plural use, i.e. the symbol for natural gas liquids is NGL not NGLs or NGL's.

Symbols	Name of Unit
atm	standard atmosphere
Bcf	billion (10^9) cubic feet
Bcm	billion (10^9) cubic metres
bbl	US barrel
b/d	barrels per day
boe (b/d oe)	barrel of oil equivalent (per day)
Btu	British thermal unit
Btu/ft³	British thermal unit per cubic foot
cal	calorie
°C	Celsius (or centigrade) degree
°F	Fahrenheit degree
ft (ft², ft³)	foot (square, cubic)
GJ	gigajoule (10^9)
GW	gigawatt (10^9)
h	hour
ha	hectare
in (in², in³)	inch (square, cubic)
J	joule
kcal	kilocalorie
kcal/h	kilocalorie per hour
kcal/kg	kilocalorie per kilogramme
kcal/m³	kilocalorie per cubic metre
kJ (kJ/kg)	kilojoule (per kilogramme)
km	kilometre
kW	kilowatt
kWh	kilowatt hour

LNG	liquefied natural gas
LPG	liquefied petroleum gas
m (m^2, m^3)	metre (square, cubic)
m^3 (st)	standard (15°C) cubic metre
mbar	millibar
Mcal	megacalorie (10^6)
MJ (MJ/m^3)	megajoule (per cubic metre)
mm	millimetre
Mcf	thousand (10^3) cubic feet
MMcf (MMcf/d)	million (10^6) cubic feet (per day)
MMBtu	million British thermal units
mrd m^3	milliard (10^9) cubic metres
MW	megawatt (10^6)
NGL	natural gas liquids
Nm^3	normal (0°C) cubic metre
scf	standard (60°F) cubic foot (30 inches mercury)
SNG	synthetic (or substitute) natural gas
syngas	synthesis gas
t	tonne
Tcf	trillion (10^{12}) cubic feet
TJ	terajoule (10^{12})
th	thermie (megacalorie 15°C value)
tce	tonne of coal equivalent
toe	tonne of oil equivalent
W	watt

SELECTED REFERENCES

The following are some of the principal references consulted by the author in compiling this book. Additional information was obtained, or consulted for verification purposes, from a wide variety of publications and journals too numerous to list.

Where apparent inconsistencies and discrepancies existed, the author has either selected what he considered to be the most reliable/accurate source of information or sought the advice of experts of the subject matter in question. For ease of reading and comprehension, no attempt has been made to litter the text with a multiplicity of references.

An Introduction to Natural Gas, Shell International Gas Ltd.
Asimov, Isaac, *A Short History of Chemistry*, Heinemann, 1972.
Beyond the Flame, Woodside Offshore Petroleum Pty. Ltd.
Blauwdruk IGU Special, N.V. Nederlandse Gasunie, 1991.
Commercial Aspects of LNG and NGL Shipping, Shell International Gas Ltd., 1984.
Comparative Safety Analysis of LNG Storage Tanks, US Department of Energy, 1982.
Cornot, Sylvie, *Underground Gas Storage in the World*, Cedigaz, 1990.
Differences between Oil and Gas – Priorities for Gas Utilisation, Shell International Gas Ltd., 1986.
Distribution Mains, The Institution of Gas Engineers, 1983.
Engineering Secure Supplies: Underground Storage, Ruhrgas A.G.
Finding Oil and Gas, Shell International Petroleum Co. Ltd., 1988.
Gainey, Brian, *Prospects for Gas in Power Generation*, Shell Coal International Ltd., 1990.
Gas Engineers Handbook, The Industrial Press.
Gas Facts in Japan 1991, Japan Gas Association.
Gaseous Fuels, edited by Louis Shnidman, American Gas Association, 1954.
Hands Across the Sea, Pertamina, 1985.
LNG Technology of Tokyo Gas, Tokyo Gas Co. Ltd., 1980.
LNG-6 Conference, various papers, 1980.
LNG-7 Conference, various papers, 1983.
LNG-8 Conference, various papers, 1986.
LNG-9 Conference, various papers, 1989.
LNG World Overview, Gotaas-Larsen Shipping Corp., 1991.
Natural Gas Transmission, N.V. Nederlandse Gasunie, 1984.
Natural Gas Transmission Across the Netherlands, N.V. Nederlandse Gasunie, 1987.
Natural Gas Terms and Measurements, Shell International Gas Ltd.
North Sea Fields Facts and Figures, Shell UK Exploration and Production.
Peebles, M.W.H., *Evolution of the Gas Industry*, The Macmillan Press Ltd., 1980.

Petroleum Handbook, 6th edition, Shell International Petroleum Co. Ltd., 1983.
Pipe Line Construction, International Pipe Line Contractors Association, 1977.
Pipeline Gas Contracts, Shell International Gas Ltd., 1984.
Reed G.E. & Ford M.E., *How Gasfields are Found and Developed*,
　Gas Engineering and Management, 1982.
Residential Gas Use, Shell International Gas Ltd., 1984.
Sean, Shell UK Exploration and Production.
Seismic Surveying, Shell International Petroleum Co. Ltd., 1990.
Shell Acid Gas Removal Processes, Shell Internationale Research Maatschappij B.V.
Technical Aspects of LNG and NGL Shipping, Shell International Gas Ltd., 1989.
16th World Gas Conference, various papers and committee reports, 1985.
17th World Gas Conference, various papers and committee reports, 1988.
18th World Gas Conference, various papers and committee reports, 1991.
World Who's Who in Science, Marquis – Who's Who Inc., 1968.